Advances in Bioceramics and Porous Ceramics IV

Advances in Bioceramics and Porous Ceramics IV

A Collection of Papers Presented at the 35th International Conference on Advanced Ceramics and Composites
January 23–28, 2011
Daytona Beach, Florida

Edited by
Roger Narayan
Paolo Colombo

Volume Editors
Sujanto Widjaja
Dileep Singh

A John Wiley & Sons, Inc., Publication

Published by John Wiley & Sons, Inc., Hoboken, New Jersey.
Published simultaneously in Canada.

Library of Congress Cataloging-in-Publication Data is available.

ISBN 978-1-118-05991-3

oBook ISBN: 978-1-118-09526-3
ePDF ISBN: 978-1-118-17264-3

ISSN: 0196-6219

10 9 8 7 6 5 4 3 2 1

Contents

Preface vii

Introduction ix

BIOCERAMICS

Fabrication of Hydroxyapatite–Calcite Nanocomposite 3
E. K. Girija, G. Suresh Kumar, A. Thamizhavel, Y. Yokogawa, and S. Narayana Kalkura

Preparation and Protein Adsorption of Silica-Based Composite Particles for Blood Purification Therapy 13
Jie Li, Yuki Shirosaki, Satoshi Hayakawa, and Akiyoshi Osaka

Collagen-Templated Sol-Gel Preparation of Ultra-Fine Silica Nanotube Mats and Osteoblastic Cell Proliferation 19
Song Chen, Toshiyuki Ikoma, Jie Li, Hiromi Morita, Akiyoshi Osaka, Masaki Takeguchi, and Nobutaka Hanagata

Tissue Ingrowth in Resorbable Porous Tissue Scaffolds 25
Janet Krevolin, James J. Liu, Adam Wallen, Kitu Patel, Rachel Dahl, Hu-Ping Hsu, Cathal Kearney, and Myron Spector

Selective Laser Sintered Ca-P/PHBV Nanocomposite Scaffolds with Sustained Release of rhBMP-2 for Bone Tissue Engineering 37
Bin Duan, William W. Lu, and Min Wang

Microbeam X-Ray Grain Averaged Residual Stress in Dental Ceramics 49
Hrishikesh A. Bale, Nobumichi Tamura, and Jay C. Hanan

Bioactive Glass Scaffolds for the Repair of Load-Bearing Bones 65
M. N. Rahaman, X. Liu, and T. S. Huang

Do Cell Culture Solutions Transform Brushite ($CaHPO_4 2H_2O$) to Octacalcium Phosphate ($Ca_8(HPO_4)_2(PO_4)_4 5H_2O$)? 79
Ibrahim Mert, Selen Mandel, and A. Cuneyt Tas

Hydroxyapatite Scaffolds for Bone Tissue Engineering with Controlled Porosity and Mechanical Strength 95
Vincenzo M. Sglavo, Marzio Piccinini, Andrea Madinelli, and Francesco Bucciotti

Hollow Hydroxyapatite Microspheres for Controlled Delivery of Proteins 101
H. Fu, M. N. Rahaman, and D. E. Day

Expression of Mineralized Tissue-Associated Proteins is Highly Upregulated in MC3T3-E1 Osteoblasts Grown on a Borosilicate Glass Substrate 111
Raina H. Jaina, Jutta Y. Marzilliera, Tia J. Kowala, Shaojie Wangb, Himanshu Jainb, and Matthias M. Falka

POROUS CERAMICS

High Porosity In Situ Catalyzed Carbon Honeycombs for Mercury Capture in Coal Fired Power Plants 123
Xinyuan Liu, Millicent K. Ruffin, Benedict Y. Johnson, and Millicent O. Owusu

Not All Microcracks are Born Equal: Thermal vs. Mechanical Microcracking in Porous Ceramics 137
Giovanni Bruno, Alexander M. Efremov, Chong An, and Seth Nickerson

SiC Foams for High Temperature Applications 153
Alberto Ortona, Sandro Gianella, and Daniele Gaia

Porous SiC Ceramic from Wood Charcoal 163
S. Manocha, Hemang Patel, and L. M. Manocha

Fabrication of Beta-Cristobalite Porous Material from Diatomite with Some Impurities 177
Osman Şan, Cem Özgür, and Remzi Gören

Microstructural Study of Alumina Porous Ceramic Produced by Reaction Bonding of Aluminium Powder Mixed with Corn Starch 185
Juliana Anggono, Ida A. O. R. S. Shavitri, and Soejono Tjitro

Characterization of Ceramic Powders during Compaction using Electrical Measurements 199
Timothy Pruyn and Rosario A. Gerhardt

Author Index 211

Preface

This issue contains the proceedings of the "Porous Ceramics: Novel Developments and Applications" and "Next Generation Bioceramics" symposia of the 35th International Conference and Exposition on Advanced Ceramics and Composites (ICACC'11), which was held on January 23–28, 2011 in Daytona Beach, Florida, USA.

A rapidly developing area of ceramic science & technology involves research into the interactions between ceramic materials and living organisms. These novel ceramic materials will provide improvements in treatment of dental and medical conditions. In addition, biomimetic ceramics, which imitate the properties and structure of materials found in nature, have generated considerable interest in the scientific community. The "Next Generation Bioceramics" symposium addressed several leading areas related to processing, characterization, and use of novel bioceramics, including advanced processing of bioceramics; biomineralization and tissue-material interactions; bioinspired and biomimetic ceramics; ceramics for drug delivery; ceramic biosensors; in vitro and in vivo characterization of bioceramics; mechanical properties of bioceramics; and nanostructured bioceramics. This symposium promoted lively interactions among various stakeholders in the bioceramics community, including researchers from industry, academia, and government.

The "Porous Ceramics" symposium brought together engineers and scientists in the area of ceramic materials containing high volume fractions of porosity, in which the porosity ranged from nano- to millimeters. These materials have attracted significant academic and industrial attention for use in different energy and environmental related applications, because of their specific properties which cannot be possessed by components in a dense form. Therefore, several contributions were devoted to the use of porous ceramics in fields such as for gas separation and particulate filtration, together with more advanced applications such as mercury capture or porous burners. Moreover, porous ceramics play a key role in medical, dental, and biotechnology applications, and the joint session involving participants from bioceramics and porous ceramics symposia enabled participants to exchange experiences and insights. Several innovations in the processing of porous ceramics were also presented, testifying to the large current scientific activity in this field.

We are convinced that this symposium will continue to play a significant role in encouraging interactions between researchers from academia and industry as well in promoting the dissemination of research in order to benefit the society at large.

We would like to thank the staff at The American Ceramic Society and John Wiley & Sons, particularly Greg Geiger, Mark Mecklenborg, Marilyn Stoltz, Marcia Stout, and Anita Lekhwani, for making this volume possible. We also give thanks to the contributors and reviewers of the proceedings volume. In addition, we thank the officers of the Engineering Ceramics Division of the American Ceramic Society and the 2011 ICACC program chair, Dr. Dileep Singh, for their support. We hope that this volume becomes a useful resource for academic, governmental, and industrial efforts involving porous ceramic and bioceramic materials. Finally, we anticipate that this volume contributes to the advancement of ceramic science and technology and signifies the leadership of The American Ceramic Society in these rapidly developing areas.

ROGER J. NARAYAN,
University of North Carolina and North Carolina State University

PAOLO COLOMBO,
Università di Padova (Italy) and The Pennsylvania State University

Introduction

This CESP issue represents papers that were submitted and approved for the proceedings of the 35th International Conference on Advanced Ceramics and Composites (ICACC), held January 23-28, 2011 in Daytona Beach, Florida. ICACC is the most prominent international meeting in the area of advanced structural, functional, and nanoscopic ceramics, composites, and other emerging ceramic materials and technologies. This prestigious conference has been organized by The American Ceramic Society's (ACerS) Engineering Ceramics Division (ECD) since 1977.

The conference was organized into the following symposia and focused sessions:

Symposium 1	Mechanical Behavior and Performance of Ceramics and Composites
Symposium 2	Advanced Ceramic Coatings for Structural, Environmental, and Functional Applications
Symposium 3	8th International Symposium on Solid Oxide Fuel Cells (SOFC): Materials, Science, and Technology
Symposium 4	Armor Ceramics
Symposium 5	Next Generation Bioceramics
Symposium 6	International Symposium on Ceramics for Electric Energy Generation, Storage, and Distribution
Symposium 7	5th International Symposium on Nanostructured Materials and Nanocomposites: Development and Applications
Symposium 8	5th International Symposium on Advanced Processing & Manufacturing Technologies (APMT) for Structural & Multifunctional Materials and Systems
Symposium 9	Porous Ceramics: Novel Developments and Applications

Symposium 10	Thermal Management Materials and Technologies
Symposium 11	Advanced Sensor Technology, Developments and Applications
Symposium 12	Materials for Extreme Environments: Ultrahigh Temperature Ceramics (UHTCs) and Nanolaminated Ternary Carbides and Nitrides (MAX Phases)
Symposium 13	Advanced Ceramics and Composites for Nuclear and Fusion Applications
Symposium 14	Advanced Materials and Technologies for Rechargeable Batteries
Focused Session 1	Geopolymers and other Inorganic Polymers
Focused Session 2	Computational Design, Modeling, Simulation and Characterization of Ceramics and Composites
Special Session	Pacific Rim Engineering Ceramics Summit

The conference proceedings are published into 9 issues of the 2011 Ceramic Engineering & Science Proceedings (CESP); Volume 32, Issues 2-10, 2011 as outlined below:

- Mechanical Properties and Performance of Engineering Ceramics and Composites VI, CESP Volume 32, Issue 2 (includes papers from Symposium 1)
- Advanced Ceramic Coatings and Materials for Extreme Environments, Volume 32, Issue 3 (includes papers from Symposia 2 and 12)
- Advances in Solid Oxide Fuel Cells VI, CESP Volume 32, Issue 4 (includes papers from Symposium 3)
- Advances in Ceramic Armor VII, CESP Volume 32, Issue 5 (includes papers from Symposium 4)
- Advances in Bioceramics and Porous Ceramics IV, CESP Volume 32, Issue 6 (includes papers from Symposia 5 and 9)
- Nanostructured Materials and Nanotechnology V, CESP Volume 32, Issue 7 (includes papers from Symposium 7)
- Advanced Processing and Manufacturing Technologies for Structural and Multifunctional Materials V, CESP Volume 32, Issue 8 (includes papers from Symposium 8)
- Ceramic Materials for Energy Applications, CESP Volume 32, Issue 9 (includes papers from Symposia 6, 13, and 14)
- Developments in Strategic Materials and Computational Design II, CESP Volume 32, Issue 10 (includes papers from Symposium 10 and 11 and from Focused Sessions 1, and 2)

The organization of the Daytona Beach meeting and the publication of these proceedings were possible thanks to the professional staff of ACerS and the tireless dedication of many ECD members. We would especially like to express our sincere thanks to the symposia organizers, session chairs, presenters and conference attendees, for their efforts and enthusiastic participation in the vibrant and cutting-edge conference.

ACerS and the ECD invite you to attend the 36th International Conference on Advanced Ceramics and Composites (http://www.ceramics.org/daytona2012) January 22-27, 2012 in Daytona Beach, Florida.

SUJANTO WIDJAJA AND DILEEP SINGH
Volume Editors
June 2011

Bioceramics

FABRICATION OF HYDROXYAPATITE–CALCITE NANOCOMPOSITE

E.K. Girija[a], G. Suresh Kumar[a], A. Thamizhavel[b], Y. Yokogawa[c], S. Narayana Kalkura[d]

[a]Department of Physics, Periyar University, Salem 636 011, India
[b]Department of Condensed Matter Physics, Tata Institute of Fundamental Research, Colaba, Mumbai 400 005, India
[c]Graduate School of Engineering, Department of Intelligent Materials Engineering, Osaka City University, Osaka 558 8585, Japan
[d]Crystal Growth Centre, Anna University, Chennai 600 025, India

ABSTRACT

Hydroxyapatite (HA) and calcite are known natural biomineral. HA is known to be bioactive and bioresorbable but the rates are too low. On the other hand, calcite is highly biodegradable. The combination of HA and $CaCO_3$ can compromise the demerits of each others. Here a method is presented for the simultaneous synthesis of HA and calcite employing sol-gel process. The product obtained was a nanocomposite with enhance bioactivity and elastic constants and fracture toughness similar to that of HA.

INTRODUCTION

Biological hard tissues are bio-composites of inorganic mineral with complex organic matrix. The inorganic minerals found in hard tissues of living organism include hydroxyapatite (HA), calcite, silica, magnetite etc.[1,2] Biomedical applications such as guided tissue regeneration, bone repair, drug delivery, tissue engineering etc requires smart biomaterial with multifunctions. There is a large demand for synthetic bone replacement materials. Materials with biocompatibility, bioactivity and bioresorbability are needed for making bone substitutes. Tissue engineering has the potential to create and regenerate tissues and organs and it requires coherent scaffold materials which can resorb at the rate of new bone regeneration.[1,2] Natural and synthetic hydroxyapatite is being investigated intensely with the aim of bone tissue engineering.[1-3] The resorption kinetics of HA is too low. Hence the biphasic mixture of HA and tricalcium phosphate has been studied by several researchers.[4] HA-calcite composite can be a more promising choice for this purpose as calcite is biocompatible, highly resorbable and can establish a direct bond with bone.[5,6] In this work, we report the simultaneous synthesis of HA-calcite nanocomposite by sol-gel method and their bioactivity and mechanical properties.

EXPERIMENTAL

Synthesis

Calcium nitrate tetrahydrate ($Ca(NO_3)_2{\cdot}4H_2O$, 98%), di-ammonium hydrogen phosphate ($(NH_4)_2HPO_4$, 99%), citric acid monohydrate ($C_6H_8O_7.H_2O$, 99.5%) and ammonia solution (NH_4OH, 25%) were used for the synthesis. The chemicals used were obtained from Merck and used without further purification. Deionized water was employed as the solvent. 1 M $Ca(NO_3)_2{\cdot}4H_2O$ and 0.6 M $(NH_4)_2HPO_4$ solutions were separately brought to pH above 10 with concentrated NH_4OH.[7] The calcium nitrate solution was stirred vigorously and di-ammonium hydrogen phosphate solution was added dropwise into it and stirred for 24 h at 60°C and the sample was named as A. The same experiment was also carried out in the presence of 1M citric acid at 60°C and the corresponding sample was referred as B. Both the as formed samples A and B were dried at 110°C for 48 h in air oven. Finally the samples were calcined at 400°C for 3 h and 1100°C for 10 h in a muffle furnace in open air atmosphere.

Characterization

XRD pattern of all synthesized and calcined samples were carried out using PANalytical X' Pert PRO diffractometer, with voltage and current setting of 40 kV and 30 mA, respectively. The XRD patterns were recorded in the range $20° \leq 2\theta \leq 60°$ at a scan speed of 1°/min giving a step size 0.0170° with Cu Kα radiation (1.5406 Å). Crystallographic identification of the phases of synthesized samples was accomplished by comparing the experimental XRD pattern with standard data compiled by the International Center for Diffraction Data (ICDD). The lattice parameters such as *a, c* and *V* were calculated by using method of least square.[8] The average particle size was calculated from XRD data using the Debye-Scherrer approximation[8]

$$D_{hkl} = \frac{K\lambda}{\beta_{1/2}\cos\theta}$$

where D_{hkl} is the particle size, as calculated for the (h k l) reflection, λ is the wavelength of CuKα radiation (1.5406 Å), $\beta_{1/2}$ is the full width at half maximum for the diffraction peak under consideration (in radian), θ is the diffraction angle (in degree) and K is the broadening constant chosen as 0.9. The diffraction peak at $2\theta = 25.8°$ was chosen for calculation of the particle size because it was sharper and isolated from others which is (002) Miller's plane of the hydroxyapatite crystal.

The degree of crystallinity (Xc) can be evaluated by the following equation[8,9]

$$X_c = \left(\frac{0.24}{\beta_{002}}\right)^3$$

where β_{002} is the full width at half maximum (degree) of (002) Miller's plane. The phase composition of synthesized samples was determined by the following equation[10]

$$RC_i = \frac{I(hkl)_i}{\sum I(hkl)_i}$$

where i and $I(hkl)_i$ refer to the phase of interest and intensity of the characteristic peaks of the corresponding phase, respectively, in the XRD pattern. $\Sigma I(hkl)_i$ is the total intensity of the characteristic peaks of the phases appeared in the XRD pattern. The HA powder samples calcined at 400°C were examined by FT–IR spectroscopy with Avatar 330 spectrometer (Thermo Nicolet, USA). The spectrum was recorded in the 4000–400 cm^{-1} region with 4 cm^{-1} resolution by using KBr pellet technique. The thermal behavior of the as-synthesized samples were determined by TG/DSC analyzer (NETZSCH, Germany) under Ar atmosphere using Al_2O_3 crucible with 20°C/min heating rate in the temperature from 25°C to 1000°C.

Bioactivity test

Bioactivity is the ability of the material to directly bond to bone through chemical interaction and not physical or mechanical attachment. Bioactivity has been characterized *in vitro* as the ability of the material to form carbonate apatite (similar to bone apatite) on its surface. The bioactivity of sample A (HA) and B (HA-calcite composite) were studied by immersing the compacted samples (pellets) in simulated body fluid (SBF) at 37°C. The SBF was prepared by dissolving appropriate amount of reagent grade NaCl, $NaHCO_3$, KCl, $Na_2HPO_4.H_2O$, $MgCl_2.6H_2O$, Na_2SO_4, $(CH_2OH)_3CNH_2$ and $CaCl_2.H_2O$ in deionized water. 1M HCl was used to maintain pH of the solution to 7.4 at 37°C to mimic the human plasma.[11] Then, the pellet samples were immersed in 30 ml of SBF in plastic containers with airtight lids and maintaining the temperature at 37°C in incubator. The SBF solution was renewed once in three days and pellet samples were immersed for a period of 21 days. The pH of the SBF solution was also measured during every renewal. Finally, the sample surface was analyzed by scanning electron microscope (SEM, JEOL-6390).

Density and porosity measurements

The density of the compacted samples were calculated using Archimede's principle by first measuring its mass, then its volume and dividing the mass by the volume. The samples were precision weighed in an electronic balance (Shimadzu Corporation, Japan) to an accuracy of 0.1 mg. The porosity of the samples was calculated by using the following relation

$$\text{Porosity} = \left(1 - \frac{\text{Measured density of sample}}{\text{Theoretical density of sample}}\right) \times 100$$

The theoretical density of samples was calculated by considering the theoretical density HA (3.16 g/cm^3) and calcite (2.71 g/cm^3) using law of mixture.[12]

Mechanical properties

We employed a non-destructive ultrasonic technique to measure the elastic constants of the samples.[13] The ultrasonic velocity (both longitudinal and shear direction) measurements were carried out using a high-power ultrasonic pulse receiver system (Fallon Ultrasonic Inc., Canada) with a 100 MHz digital storage oscilloscope (HP-54600B) using the cross-correlation technique described elsewhere.[13] Precise transit time for the propagation of the ultrasonic waves into the samples was measured by taking the difference in the transit times (Δt) between t_1 and t_2, where t_1 is the transit time measured only with buffer rods at a given temperature and t_2 is the transit time measured after introducing the sample in between the buffer rods, at the same temperature by using the following relation

$$U = \frac{\Delta t}{d}$$

where d is the thickness of samples. Elastic (longitudinal (L), shear (G) and Young's (E)) moduli and Poisson's ratio (σ) of the samples was calculated from the measured ultrasonic velocities (U_L and U_S) and density using the following relations. [13]

$$\text{Longitudinal modulus, } L = U_L^2\rho$$

$$\text{Shear modulus, } G = U_S^2\rho$$

$$\text{Poisson's ratio, } \sigma = \frac{(L-2G)}{2(L-G)}$$

$$\text{Young's modulus, } E = (1+\sigma)2G$$

To determine the Vicker's hardness value, compacted samples (pellets) were subjected to a load of 200 g for 15 s with a Vicker's indenter (Micro Hardness Tester, HMV-2 Series, Shimadzu Corporation, Japan). A total of 3 indentations were made and the hardness values were averaged and reported. Vicker's hardness was calculated by using the following equation[14]

$$\text{Vicker's hardness, } H_V = 1.844 \times \frac{P}{d^2}$$

where P –Load (Kgf), d- Length of the diagonal of Vicker's indent. The fracture toughness K_C of the samples can be calculated using the following equation[14]

$$\text{Fracture toughness } K_C = 0.016\left(\frac{E}{H}\right)^{0.5}\left(\frac{P}{C^{1.5}}\right)$$

where E – Young's modulus (GPa), H – Vicker's hardness (GPa), P - Load (Kgf), C - Sum of half diagonal of Vicker's indent and crack length emanating from the corner of the Vicker's indent (mm).

RESULTS AND DICUSSIONS

XRD patterns of the as-synthesized gel powders are given in Fig. 1. Comparison of these patterns with JCPDS files conformed that the sample A is HA (JCPDS file No. 09–0432) with ammonium nitrate (JCPDS file No. 85–1093) and sample B contains minor quantity of β–TCP (JCPDS file No. 09–0169) and ammonium nitrate.

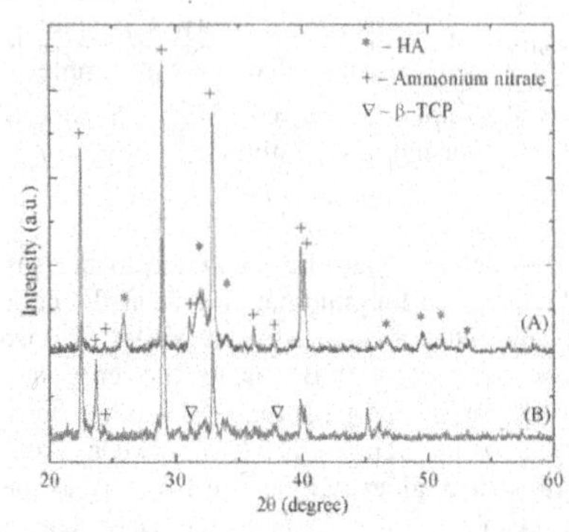

Fig.1. XRD patterns of the as-synthesized powders.

The XRD patterns of 400°C calcined samples showed the presence of pure HA in A and HA with an additional phase calcite (JCPDS file No. 47–1743) in B. The phase composition of HA and calcite in sample B was 76% and 22% respectively. When citric acid was introduced into the reaction medium it chelated with Ca ions and form Ca-citrate complex. Since the working pH was above 8 Ca-citrate formations was not much favored.[15] Only less amount of citrate ions complexed with Ca ions. The condition prevailed was favorable for the calcium phosphate formation. Hence in the as prepared sample trace amount of β–TCP existed and rest of the calcium phosphate may be in the amorphous form. Poorly crystalline HA phase has evolved along with calcite on 400°C calcination. The less amount of Ca-citrate existed in the as-synthesized powder has reacted with atmospheric oxygen and formed calcite according to the following reaction[16]

$$(C_6H_5O_7)_2Ca_3 + 9O_2 \quad 3CaCO_3 + 9CO_2 + 5H_2O$$

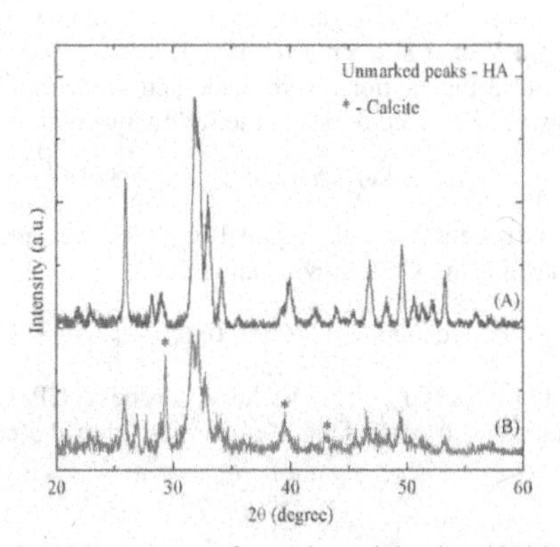

Fig.2. XRD patterns of samples calcined at 400°C.

Table 1 The crystalline parameters of HA powder calcined at 400°C.

Sample code	Lattice parameter (Å)		Average crystallite size D_{hkl} (nm)	Unit cell volume V (Å^3)	Lattice distortion c/a	Degree of crystallinity X_C
	$a = b$	c				
A	9.4238	6.8764	84	528.86	0.7296	15.6
B	9.4285	6.8808	55	529.73	0.7297	04.3

Table 1 gives the calculated values of crystalline parameters of samples calcined at 400°C. Existence of citric acid slightly increased the lattice parameters of HA. Crystallinity and average crystallite size was significantly reduced when citric acid was added during the synthesis. The free citrate ions existed in the medium might adsorb onto the precipitated amorphous calcium phosphate particles and would have controlled the size of particles. This might be the reason for the reduced crystllite size than the control samples.[17]

XRD patterns of samples calcined at 1100°C are shown in Fig. 3. On calcination at 1100°C HA has partially decomposed into oxyapatite (JCPDS file No. 89–6495) and β–TCP. Calcite has decomposed into CaO (JCPDS file No. 04–0777). The Phase composition of both samples calcined at 1100°C is shown in Table 2. The other phases such as oxyapatite, β–TCP and CaO formed when calcined at 1100°C indicate the thermal instability of the products synthesized.

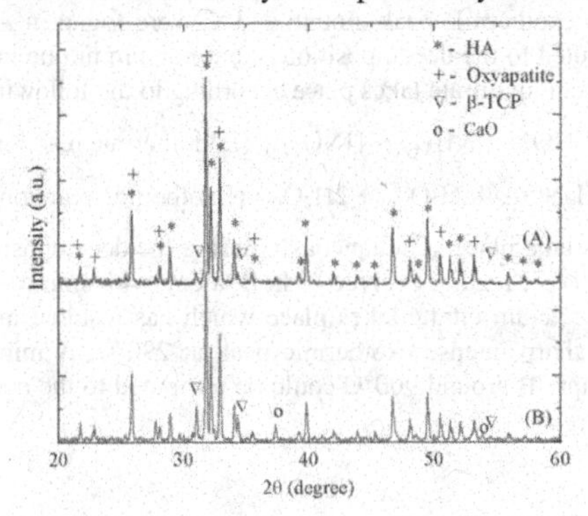

Fig. 3. XRD patterns of samples calcined at 1100°C.

Table 2 Phase composition of both samples calcined at 1100°C.

Sample code	Phase composition (%)			
	HA	Oxyapatite	β–TCP	CaO
A	49.3	49.3	1.4	-
B	41.7	41.7	11.3	5.3

The FT–IR spectra of samples calcined at 400°C are shown in the Fig. 4. All the four different vibrational modes of PO_4^{3-} were observed in the FT-IR spectra. The characteristic v_4 peaks of HA present at 575–610 cm^{-1} and the characteristic vibrational modes of OH group of HA at 3570 cm^{-1} and 628 cm^{-1} confirmed the phase formed to be HA.[6,7,18] The peaks present at 875 cm^{-1}, 1415 cm^{-1} and 1459 cm^{-1} in FT–IR spectra of samples A and B are attributed to the CO_3^{2-} ions. The FT-IR spectra of sample B showed all the above mentioned peaks but the CO_3^{2-} ion absorption peaks were stronger and wider than those for the sample A suggesting that there are more CO_3^{2-} groups present in these samples. The existence of lattice water in both the samples was confirmed from the strong band at 1630–1636 cm^{-1} and a broad band between 3550–3200 cm^{-1}.[6, 7, 18]

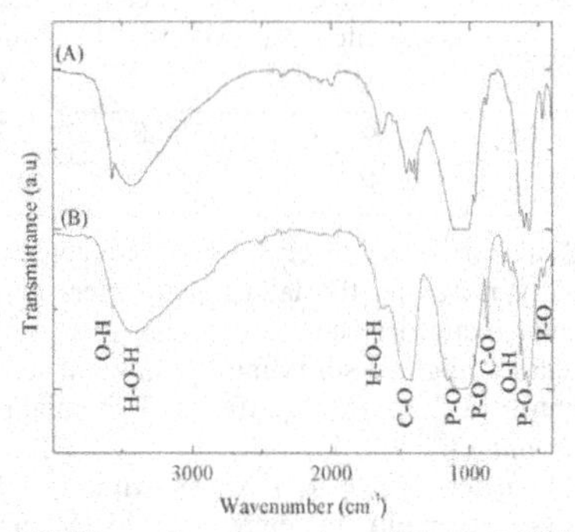

Fig. 4. FT–IR spectra of samples calcined at 400°C.

Fig. 5 shows the DSC trace of both samples. The endothermic peak around 100°C observed in A and B is due to the evaporation of adsorbed water.[6,7] With increasing temperature, an endothermic peak around 310°C and an exothermic peak around 313°C were found in A. These endothermic and exothermic peaks are attributed to the decomposition of ammonium nitrate in the as-prepared samples. The decomposition of ammonium nitrate takes place according to the following reactions

$$NH_4NO_3 \quad NH_{3\,(g)} + HNO_{3\,(g)} \text{ (Endothermic reaction)}$$

$$NH_4NO_3 \quad N_2O_{\,(g)} + 2H_2O_{(g)} \text{ (Exothermic reaction)}$$

These two reactions take place simultaneously during the decomposition of ammonium nitrate and the gaseous byproducts were readily evolved.[6,7] In B, along with ammonium nitrate decomposition of residual citric acid and calcium citrate takes place which has resulted in broad endothermic peak around 230°C and a very sharp intense exothermic peak at 290°C. A minute endothermic fall that occurs in DSC trace of sample B around 900°C could be attributed to the decorboxylation of samples, releasing CO_2.

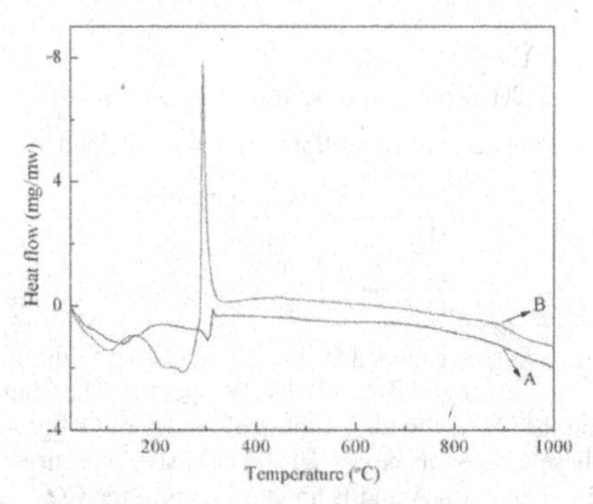

Fig. 5. DSC curve of as-synthesized gel powders.

TG curve of as-prepared samples is shown in Fig. 6. Both samples showed approximately 4–6% weight loss around 100°C, which is due to desorption of adsorbed water.[6,7] With increasing temperature, a weight loss of about 55% for sample A was observed between 100 and 300°C, which corresponded to the decomposition of ammonium nitrate. In case of samples B, a weight loss of about 70% was observed between 100 and 300°C which are attributed to decomposition of citric acid and ammonium nitrate. A weight loss about 10% was observed in between 280°C and 1000°C may be due to the slow release of CO_2.

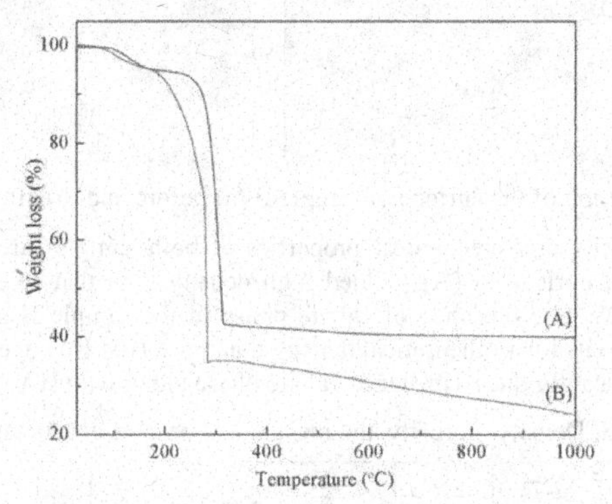

Fig. 6. TG curve of as-synthesized gel powders.

Fig. 7 and 8 shows the SEM images of the surfaces of sample A (HA) and sample B (HA-calcite composite) before and after soaking in SBF for 21 days. The surface of HA after soaking in SBF had shown the formation of spherical apatite particles of diameter >1μm on the surface. But the number of spherical deposits formed and the surface coverage was less. On the other hand, the surface of HA-calcite composite formed large number of clusters of spherical deposits and the surface was largely covered with such deposits. The inset in the figures shows the magnified view of the spherical deposit which depicts the flaky nano HA crystals constituting the spherical deposits. The pH of the SBF solution decreased during immersion of the pellets and the decrease was more for the HA-calcite composite. The consumption of OH ions from the solution to form the apatite on the surface of the pellet is the reason for the reduction in pH. The high bioactive nature of the HA-calcite composite is obvious from this study.

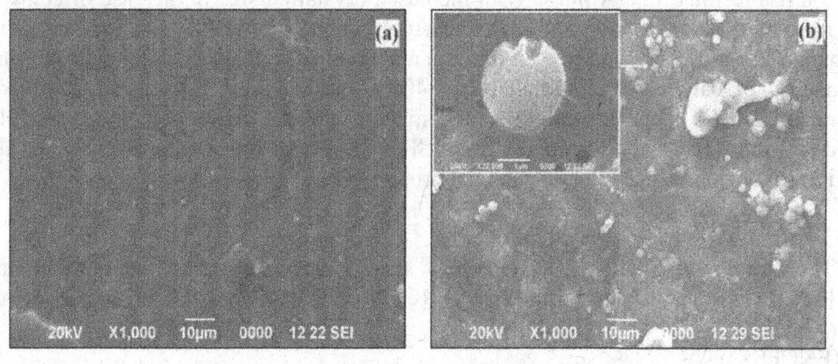

Fig. 7. SEM photograph of the surface of sample A (a) before and (b) after soaking in SBF.

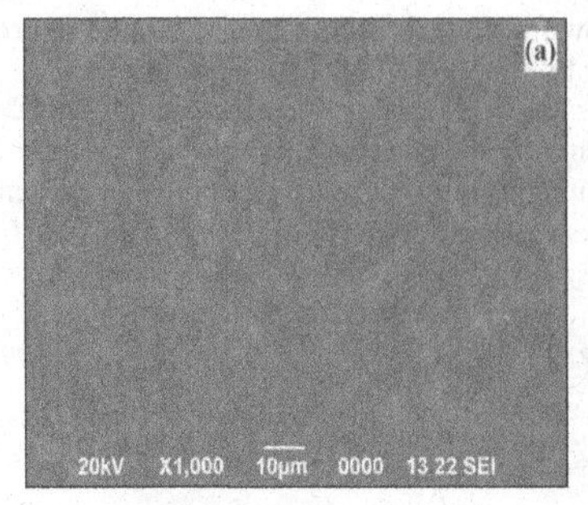

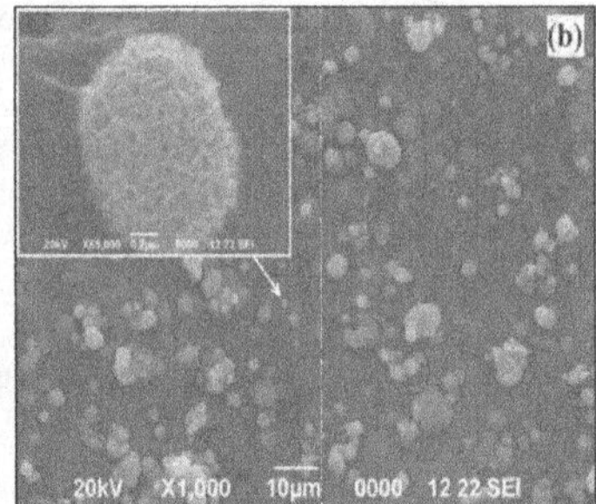

Fig. 8. SEM photograph of the surface of sample B (a) before and (b) after soaking in SBF.

The density, porosity and mechanical properties of both samples are given in Table 3. The samples prepared without citric acid (A) yielded high density than that of samples prepared in the presence of citric acid (B). The presence of calcite phase in the sample B significantly reduces the density of sample B due to its lower theoretical density than pure HA. However, mechanical properties did not get affected significantly due to presence calcite phase along with HA.

Table 3. Density, porosity and mechanical properties of samples.

Sample code	Density, (g/cm^3)	Porosity (%)	Longitudinal velocity, U_L (ms^{-1})	Shear velocity, U_S (ms^{-1})	Longitudinal modulus, L (GPa)	Shear modulus, G (GPa)	Poisson's ratio,	Young's modulus, E (GPa)	Hardness, H (GPa)	Fracture toughness, K_C ($MPa.m^{1/2}$)
A	2.30	27	2729	1654	17.12	06.29	0.209	15.20	1.029	0.37
B	2.27	22	2721	1673	16.80	06.35	0.196	15.18	0.637	0.38

CONCLUSIONS

Colloidal mixture of calcium and phosphate precursor followed by 110°C drying and 400°C calcination has resulted in HA phase with increased crystallite size in the case of control experiments and a nanocomposite of HA and calcite with reduced crystallite size in the citric acid used experiments. The citric acid introduced into the reaction medium has chelated with Ca ions resulting in the formation of calcite along with HA on 400°C calcinations. Citric acid has also reduced the crystallite size of HA. This HA–calcite nanocomposite showed higher bioactivity than HA which may due to dissolution of Ca ions from calcite in SBF. However, the presence of calcite along with HA neither improved nor reduced the mechanical properties.

ACKNOWLEDGMENTS

The authors (G.S and E.K.G) express their sincere thanks to Department of Science and Technology (DST), New Delhi, India (Project Ref. No: SR/FTP/PS-24/2009) for financial support.

REFERENCES

[1]S.V. Dorozhkin, Bioceramics of calcium orthophosphates, *Biomaterials*, **31**, 1465-1485 (2010).

[2]S.J. Kalita, A. Bhardwaj, H.A. Bhatt, Nanocrystalline calcium phosphate ceramics in biomedical engineering, *Mater. Sci. Eng. C*, **27**, 441-449 (2007).
[3]W. Suchanek, M. Yoshimura, Processing and properties of hydroxyapatite-based biomaterials for use as hard tissue replacement implants, *J. Mater. Res.*, **13**, 94-116 (1998).
[4]R.Z. LeGeros, S. Lin, R. Rohanizadeh, D. Mijares, J.P. LeGeros, Biphasic calcium phosphate bioceramics: preparation, properties and applications, *J. Mater. Sci.: Mater. Med.*, **14**, 201-209 (2003).
[5]Y. Fujita, T. Yamamuro, T. Nakamura, S. Kotani, The bonding behavior of calcite to bone, *J. Biomed. Mater. Res.*, **25**, 991-1003 (1991).
[6]G.S. Kumar, E.K. Girija, A. Thamizhavel, Y. Yokogawa, S.N. Kalkura, Synthesis and characterization of bioactive hydroxyapatite-calcite nanocomposite for biomedical applications, *J. Colloid Interface Sci.*, **349**, 56-62 (2010).
[7]E. Hayek and H. Newesely, Pentacalcium Monohydroxyorthophosphate, *Inorg. Synth.*, 7, 63-65 (1963)
[8]F. Ren, R. Xin, X. Ge, Y. Leng, Characterization and structural analysis of zinc-substituted hydroxyapatites, *Acta Biomater.*, **5**, 3141-3149 (2009).
[9]E. Landi, A. Tampieri, G. Celotti, S. Sprio, Densification behaviour and mechanisms of synthetic hydroxyapatites, *J. Eur. Ceram. Soc.*, **20**, 2377-2387 (2000).
[10]C. Ergun, Effect of Ti ion substitution on the structure of hydroxylapatite, *J. Eur. Ceram. Soc.*, **28**, 2137-2149 (2008).
[11]A.C. Tas, Synthesis of biomimetic Ca-hydroxyapatite powders at 37°C in synthetic body fluids, *Biomaterials*, **21**, 1429-1438 (2000).
[12]Y. Hu, X. Miao, Comparison of hydroxyapatite ceramics and hydroxyapatite/borosilicate glass composites prepared by slip casting, *Ceram. Int.*, **30**, 1787-1791(2004).
[13]O. Prokopiev, I. Sevostianov, Dependence of the mechanical properties of sintered hydroxyapatite on the sintering temperature, *Mater. Sci. Eng. A*, **431**, 218-227 (2006).
[14]J. Wang, L.L. Shaw, Nanocrystalline hydroxyapatite with simultaneous enhancements in hardness and toughness, *Biomaterrials*, **30**, 6565-6572 (2009).
[15]M.A. Martins, C. Santos, M.M. Almeida, M.E.V. Costa, Hydroxyapatite micro- and nanoparticles: Nucleation and growth mechanisms in the presence of citrate species, *J. Colloid Interface Sci.*, **318**, 210-216 (2008).
[16]A.I. Mitsionis, T.C. Vaimakis, C.C. Trapalis, The effect of citric acid on the sintering of calcium phosphate bioceramics, *Ceram. Int.*, **36**, 623-634 (2010).
[17]C. L. Chu, P.H. Lin, Y. S. Dong, D.Y. Guo, Influences of citric acid as a chelating reagent on the characteristics of nanophase hydroxyapatite powders fabricated by a sol-gel method, *J. Mater. Sci. Lett.*, **21**, 1793-1795 (2002).
[18]S. Koutsopoulos, Synthesis and characterization of hydroxyapatite crystals: A review study on the the analytical methods, *J. Biomed. Mater. Res.*, **62**, 600-612 (2002).

PREPARATION AND PROTEIN ADSORPTION OF SILICA-BASED COMPOSITE PARTICLES FOR BLOOD PURIFICATION THERAPY

Jie Li, Yuki Shirosaki, Satoshi Hayakawa, Akiyoshi Osaka*
The Graduate School of Natural Science and Technology, Okayama University
Tsushima, Kita-ku, Okayama-shi 700-8530, Japan
* a-osaka@cc.okayama-u.ac.jp

ABSTRACT

Silica gel macrospheres of a few millimeters in diameter were fabricated via the sol-gel route using water glass, calcium chloride, and sodium alginate as the precursor components. TiO_2 coating on the silica gel macrosphere surface was conducted by soaking them in titania sol derived from hydrolysis of tetraethylorthotitanate (TEOT) under varied pH (HNO_3), and temperature (50~80°C). X-ray diffraction analysis indicated deposition of anatase, and particular deposition (<1μm) on the surface was confirmed by scanning electron microscopy. Protein adsorption behavior was examined as a function of pH of the hydrolysis of TEOT, using bovine serum albumin (BSA) and lysozyme (LSZ) as the model proteins. BSA adsorption was little affected by the pH, while some effects were found for LSZ adsorption. BSA adsorption saturated at 20~25 min contact, while LSZ adsorption took longer time for the saturation. Moreover, BSA was adsorbed almost twice as much as LSZ, when saturation adsorption was compared. An effect of residual calcium ions with better affinity to BSA was proposed for partial interpretation since surface charge could not account for those adsorption behaviors.

INTRODUCTION

A variety of biomaterials with high affinity to proteins have attracted attention recently as good candidates for human blood purification materials. Current blood purification therapy employs mostly organic components, such as pHEMA-coated active carbon and resins such as 2-(diethylamino)ethyl methacrylate-co-ethylene glycol dimethacrylate polymers[1], and dextran- and polyethylene glycol-modified polydimethylsiloxane[2]. Very few proposed inorganic or ceramic protein adsorbents: Zn- and carbonate ion-containing hydroxyapatite[3-5] or sol-gel derived TiO_2[6,7], while several studies were carried out on blood compatibility of titania[8-12]. Most of the titania studies[8-12], however, were on titania layers on metallic implants or sheet of Ti, and did not take blood purification devices into any account. At this moment, it is almost impossible to shape titania into hollow and fibrils like those for dialysis, and hence, coating of biologically active titania nano- or microparticles is to be achieved on some spherical granules as the substrates when adsorption is a phenomenon associated with the surface. Therefore, the choice of the ceramic substrate and control of surface morphology are principal key factors. Takemoto et al. [12] and Asano et al. [6,7] pointed out that anatase particles were highly blood compatible and active against specific pathogenic proteins, showing moderate affinity to albumin. Moreover, Ti-OH groups are active under aqueous conditions, and then, those titania may be favorably derived from aqueous sol-gel systems. As far as sol-gel procedure is concerned, silica macro-spheres are the best selection for the substrate because the resultant silica gel spheres inevitably involve silanol groups that are also chemically active to fix titania particle on their surface. When the adsorbents are thrown away after use, or even if they are reactivated via the cleaning process, commercialization as columns for blood perfusion therapy or other demands to suppress the fabrication cost as low as possible. This rationalized the use of water glass as it is a low cost source of Si(IV).

In consequence, composite particles of a silica core and a biologically active titania shell with an adequate size distribution are more favorable than those films, membranes, and particulates materials. In this study, millimeter-size silica macrospheres with TiO_2 coating were prepared via the sol-gel route using water glass and tetraethylorthotitanate (TEOT; $Ti(OC_2H_5)_4$) as the silica and TiO_2 precursors, respectively. Sodium alginate and calcium chloride were employed for yielding the silica gel spheres. Titanium oxide colloidal particles were coated onto the silica macrospheres. These gel spheres were characterized with using X-ray diffractometry (XRD), scanning electron microscopy (SEM), and inductively coupled plasma photometry (ICP). Protein adsorption on those macrospheres was examined, using bovine serum albumin (BSA) and lysozyme (LSZ) as the model proteins.

EXPERIMENTAL

Preparation of silica gel macrospheres coated with anatase particles

Figure 1 schematically demonstrates the fabrication procedure of the silica gel macrospheres. Water glass (55mass%; nacalai tesque, Osaka, Japan) was first diluted to 30 mass % aqueous solution, and was drop-wisely added to 1M HCl solution under vigorous stirring at 30°C until pH of the HCl solution reached 3.5. The obtained silica sol was mixed with 1.5 mass% sodium alginate solution (pH=7.2) in mass ratio of 3:2 under stirring at 30 °C. This mixture was metastable, and became gel or peptized when kept ~3 d at room temperature. This freshly prepared silica sol was added dropwisely to 3.5 mass% $CaCl_2$ (pH 6.4) aqueous solution at room temperature, using syringe (=1.2mm) feeder to yield silica gel macrospheres. They were aged in the $CaCl_2$ solution for 3d at room temperature, and rinsed in deionized water with ultrasonic bath with 5min for 3 times to remove the sodium ions introduced to the systems associated with the water glass.

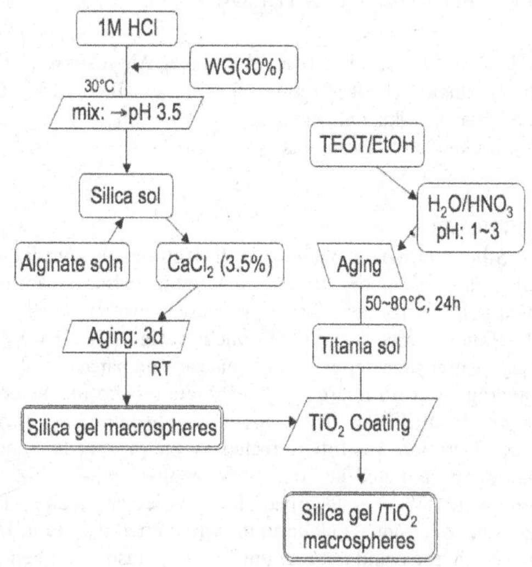

Fig. 1 The flow chart for fabricating silica gel macrospheres and anatase-coated silica macrospheres. WG: water glass; Alg: Na alginate; TEOET: $Ti(OC_2H_5)_4$; EtOH: $C_2H_5)OH$. mass%.

TiO_2 Coating on Silica Particles

Dry ethanol (EtOH), deionized water and TEOT were employed in this experiment so that the molar ratio 5 : 128 : 1 was attained. EtOH was mixed with TEOT under vigorous stirring. The resultant solution was added to H_2O at room temperature, pH of which was adjusted to 1~5 with HNO_3 (99%). The transparent solution was aged at 50° ~ 80°C for 0-24h to yield titania sol. Then, about 20 silica macrospheres were soaked in the TiO_2 sol (30mL), aged for ~24h at room temperature for titania coating, and then rinsed by deionized water for 3 times, before these particles were dried in thermostat at 40 °C overnight.

Protein Adsorption

The protein adsorption measurement was conducted following an established procedure[4], bovine serum albumin (BSA) and lysozyme (LSZ) were dissolved in 30 ml of saline, with pH adjusted to 7.4 with Tris buffer, to prepare 35mg/ml solution of each protein. After soaking 20 pieces of titania-coated silica macrospheres for 0-120min, the amount of residual proteins in the solutions was measured photometrically, employing the optical adsorption at 280nm for BSA and 320 nm for LSZ.

RESULTS

Silica macrospheres of <3 mm in diameter were obtained, while the size increased with the rate of feeding silica sol. When pH of the water to hydrolyze TEOT exceeded 4, the titania sol system became gel. Thus, the hydrolysis of TEOT was conducted under pH 1~3. The state of resultant titania sol, crystallinity, particle size, or hydrolysis and condensation are naturally dependent on the hydrolyzing pH, temperature, or aging period in the procedure. Fig. 2 shows the XRD patterns of the titania-coated silica macrospheres where the diffractions are all assignable to anatase. Fig. 2(a) shows the effects of pH under which TEOT was hydrolyzed, (b) shows those of hydrolyzing temperature (pH = 2), and (c) shows those of hydrolyzing period (0~24 h; pH = 2). Fig. 2(a) indicates no different effects of changing pH of the hydrolyzing solution on the XRD profiles. In Fig. 2(b), the higher temperature yielded a little stronger diffraction peaks. Note the absence of XRD profile for 80°C in Fig. 2(b): it was also found that the coating sol system became gel on keeping at 80°C for 24 h. Fig. 2(c) indicates that the XRD intensity increased up to 3h, while longer hydrolysis than 24h gave no significant improvement. Our goal is to design appropriate silica macrospheres whose protein adsorption characteristics are optimized. Thus, the titania layer silica macrospheres should not necessarily have better crystallinity or stronger XRD intensity of titania. Yet, stronger diffraction intensity means a larger amount of anatase coated on the silica gel spheres, and such macrospheres may lead exhibit better adsorption characteristics.

The SEM images in Fig. 3 show significantly different surface morphology after the anatase coating

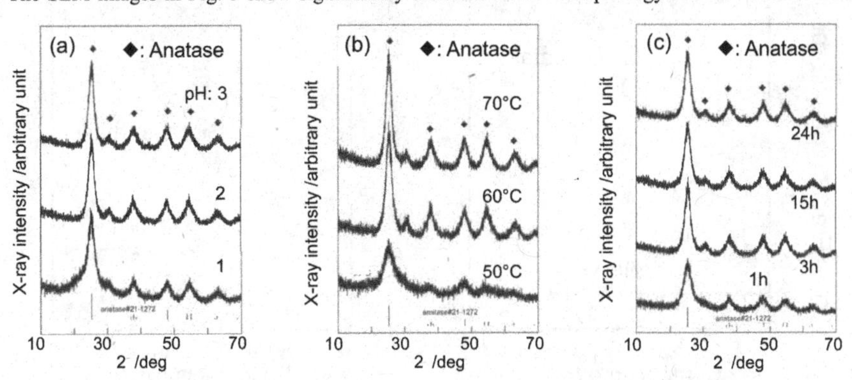

Fig. 2 XRD patterns of the titania-coated silica gel macrospheres. The XRD peaks are all assignable to anatase. The effects of (a) pH of the hydrolyzing solution (70°C, 24 h), (b) temperature of hydrolysis (pH =2, 24h), and (c) hydrolysis period (70°C, pH = 2) are shown.

procedure. The surface of the original silica gel macrosphere (Fig. 3(a)) seems rather smooth, and the anatase coating yielded rougher surface with many semi-spherical particles (Fig. 3(b)). From the XRD profiles in Fig. 2 (c), the deposited particles should be anatase.

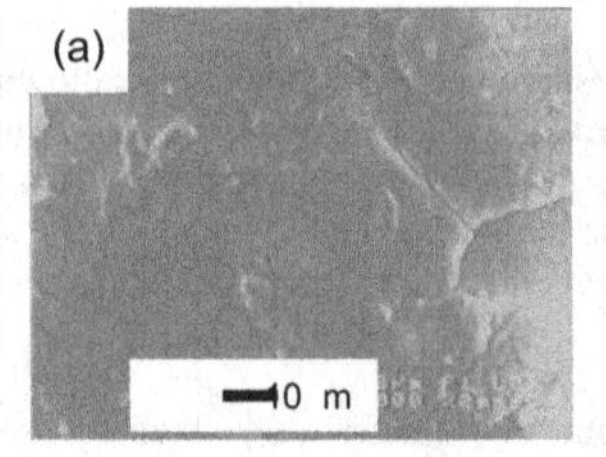

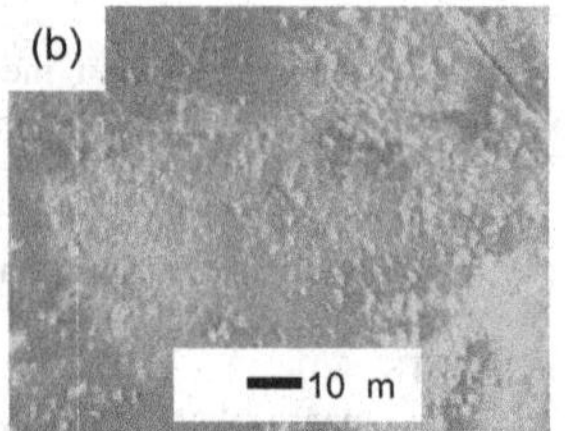

Fig. 3 Surface morphology of the silica gel macrospheres: (a) original macrosphere; (b) the anatase-coated macrosphere. Titania sol: hydrolysis at 70 °C for 3h under pH 2.

From the results above, the macrospheres that were soaked in the titania sol derived from hydrolysis under pH = 1~3, and at 70°C for 3h, were employed for the protein adsorption assessment.

Fig. 4 shows profiles of BSA and LSZ adsorption on the silica gel macrospheres with and without anatase coating. Here, the fraction of BSA and LSZ (mass %), adsorbed on 20 pieces of macrospheres with 3 mm in average diameter, was plotted as a function of the contacting time. BSA adsorption increased quickly up to 20 ~ 25 min until it reached plateau stages or saturation; that is, no more BSA was adsorbed with longer contact time. LSZ adsorption showed similar profiles, while the rate of adsorption for each

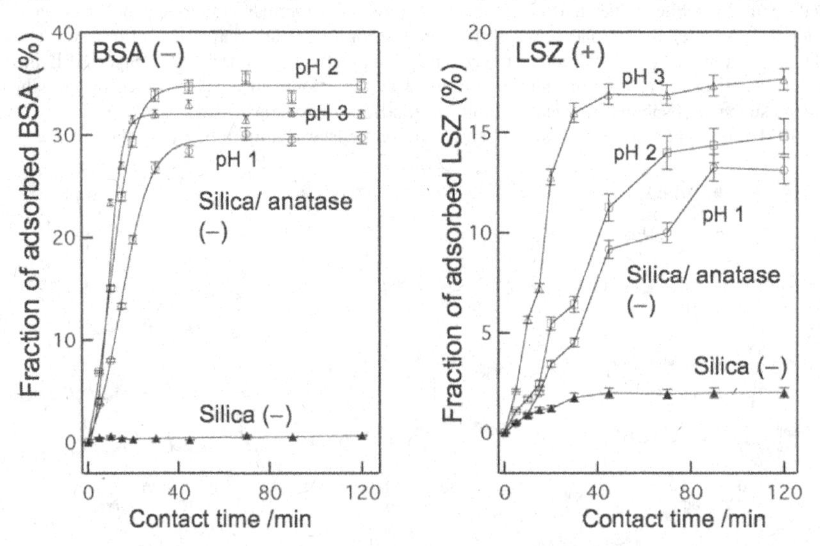

Fig. 4. Profiles of BSA (left) and LSZ adsorption on 20 silica gel macrospheres, as a function of the contact time. Fraction in mass%. pH for hydrolysis of TEOT is indicated. Hydrolysis: 70 °C, 3h; coating: RT, 24 h. Negative (-) and positive (+) charges residing on each species are indicated.

macrosphere at the earlier stage before reaching saturation was smaller than that for BSA. Moreover, the amount for the saturated adsorption of BSA was greater than that of LSZ, or the present macrospheres had better affinity to BSA than LSZ, except the anatase-free silica gel macrospheres that adsorbed more LSZ than BSA. After contacting longer than 40 min, the saturation adsorption of BSA was greatest for the anatase-coated macrospheres derived from the pH 2 titania sol, followed by those from pH 3, and then pH 1. In contrast, the greatest saturation adsorption of LSZ was found for those from the pH 3 sol and the least was for those from pH 1.

DISCUSSION

The adsorption profiles are the functions of pH of the medium, surface charge of the adsorbent (point of zero charge (pzc), or zeta potential), and surface topography.[9,13] The surface topography explains partly the difference between the saturation adsorption on the macrospheres with and without the anatase coating, but does not apply to the difference among the anatase-coated macrospheres because those had surface microstructures similar to each other. As pzc for BSA (pH 4.7) and LSZ (pH ~11) is in the acidic and basic range, respectively, each protein was charged negatively and positively in an aqueous medium of pH 7.4 as in the present study. Both anatase (~6.8 in pzc) and silica gel (~2 in pzc) are negatively charged. Then, positively charged LSZ may induce much better affinity to negatively charged surface than BSA. Thus, if surface charge effects work perfectly, the saturation adsorption of BSA and LSZ must be in the order LSZ >>BSA. This is just the inverse order found in Fig. 4. Moreover, such average surface charge model would not be able to interpret the fact that BSA was adsorbed almost twice as much LSZ. Thus, considering both negative and positive charges to be distributed on the protein molecules, an average charge on the macrospheres cannot be a major factor that controls the protein adsorption.

A possible difference in BSA and LSZ is affinity to cations like Na^+ and Ca^{2+}: BSA was better than LSZ. The starting agents in this study involved much sodium ions, and calcium ions were forced to chelate with alginate. The macrospheres were subjected to washing so that the content of free Ca(II) in the macrospheres was suppressed to a level at an equilibrium with the rinsing water of 0.6 mM Ca(II), much lower than the Ca(II) concentration in the blood plasma (2.5 mM). Yet, some of Ca(II) are left in the macrosphere, such calcium ion sites are one of the target sites of BSA adsorption.

CONCLUSION

Silica macrospheres of 3mm in average size were prepared via a sol-gel route, using water glass, sodium alginate and calcium chloride. Tetraethylorthotitanate was hydrolyzed in the solvent ethanol-water with HNO_3 as the catalyst under varied pH (1~5) at temperatures 50° to 80°C. The silica gel macrospheres were soaked in the resultant titania sol and was coated with anatase. X-ray diffraction analysis indicated the presence of anatase, and scanning electron micrographs confirmed spherical depositions on the silica gel macrospheres. The macrospheres, coated with anatase using the titania sol derived from hydrolyzing at 70°C at pH 1~3 for 3h, were employed for protein adsorption assessment. Adsorption of BSA was little affected by the pH of hydrolysis for the titania sol, while some difference was found for LSZ adsorption. Moreover, BSA was adsorbed almost twice as much as LSZ, when saturation adsorption was compared. In addition, BSA adsorption reached the saturation at 20~25 min contact, while LSZ adsorption took longer time for reaching the saturation. The surface charge could not account for those adsorption behaviors. The residual calcium ions on the surface were proposed to be one of the factors to interpret the better affinity of the macrospheres to BSA.

ACKNOWLEDGMENT

Partial financial support by Wesco Scientific Promotion Foundation (Japan) is acknowledged.

REFERENCES

[1]Barral S, Guerrelro A, Vilia-Garcia M A, Rendueles M, Diaz M, Piletsky S., Synthesis of 2-(diethylamino)ethyl methacrylate-based polymers: effect of crosslinking degree, porogen and solvent on the textural properties and protein adsorption performance, *Reactive and Functional Polymers*, **11**, 890-899 (2010).

[2]Farrell M, Beaudoin S. Surrface forces and protein adsorption on dextran- and polyethylene glycol-modified polydimethylsiloxane, *Colloids and Surfaces B: Biointerfaces*, **2**, 468-475 (2010).

[3]Fujii E, Ohkubo M, Tsuru K, Hayakawa S, Osaka A, Kawabata K, Bonhomme C, Babonneau F., Selective protein adsorption property and characterization of nano-crystalline Zinc-coating hydroxyapatite, *Acta Biomaterialia*, 2, 69-74 (2006).

[4]S. Takemoto, Y. Kusudo, K. Tsuru, S. Hayakawa, A. Osaka, S. Takashima, Selective protein adsorption and blood compatibility of hydroxyl-carbonate apatites, *J. Biomed. Mat. Res. Part A*, **69A,** 544-551 (2004).

[5]Kandori K, Shimizu T, Yasukawa A, |Shikawa T, Adsorption of bovine serum albumin onto synthetic calcium hydroxyapatite: influence of particle texture, *Colloids and Surfaces B. Biointerfaces*, **5**, 81-87 (1995).

[6]T. Asano, S. Takemoto, K. Tsuru, S. Hayakawa, A. Osaka, S. Takashima, Sol-gel preparation of blood-compatible titania as an adsorbent of bilirubin, *J. Ceram. Soc. Japan*, **111**, 645-650 (2003).

[7]T. Asano, K. Tsuru, S. Hayakawa, A. Osaka, Bilirubin adsorption property of sol-gel-derived titania particles for blood purification therapy, *Acta Biomaterialia*, 4,1067-1072 (2008).

[8]B. Wälivarra, B-O. Aronsson, M. Rodahl, J. Lausmaa, P. Tengvall, Titanium with different oxides: in vitro studies of protein adsorption and contact activation, *Biomaterials*, **15**, 827-834 (1994).

[9]JY. Park, CH. Gemmell, JE. Davies, Platelet interactions with titanium: modulation of platelet activity by surface topography, *Biomaterials*, **22**, 2671-2682 (2001).

[10]C. Eriksson, J. Lausmaa and H. Nygren, Interactions between human whole blood and modified TiO_2-surfaces: influence of surface topography and oxide thickness on leukocyte adhesion and activation, *Biomaterials*, **22**, 1987-1996 (2001)

[11]N. Huang, P. Yang, X. Cheng, Y-X. Leng, X-L. Zheng, G-G. Cai, Z-H. Zhen, F. Zhang, Y-R. Chen, X-H. Liu, T-F. Xi Tingfei, Blood compatibility of amorphous titanium oxide films synthesized by ion beam enhanced deposition, *Biomaterials*, **19**, 771-776, (1998).

[12]S. Takemoto, K. Tsuru, S. Hayakawa, A. Osaka, S. Takashima, Highly Blood Compatible Titania Gel, *J. Sol-Gel Sci. Technol.*, **21**, 97-104 (2001).

[13]S. R. Sousa, M. Lamghari, P. Sampaio, P. Moradas-Ferreira5, M. A. Barbosa, Osteoblast adhesion and morphology on TiO_2 depends on the competitive preadsorption of albumin and fibronectin, *J. Biomed. Mat. Res. Part A*, **84A**, 281-290 (2008).

COLLAGEN-TEMPLATED SOL-GEL PREPARATION OF ULTRA-FINE SILICA NANOTUBE MATS AND OSTEOBLASTIC CELL PROLIFERATION

Song Chen,[a] Toshiyuki Ikoma,[b] Jie Li,[c] Hiromi Morita,[c] Akiyoshi Osaka,[d] Masaki Takeguchi,[e] and Nobutaka Hanagata[c,f]

[a] Biomaterials Center, National Institute for Materials Science, Sengen, Tsukuba, 305-0047, Japan
[b] Department of Metallurgy and Ceramics Science, Tokyo Institute of Technology, Meguro-ku, Tokyo 152-8550, Japan
[c] Nanotechnology Innovation Center, National Institute for Materials Science, Sengen, Tsukuba 305-0047, Japan
[d] Graduate School of Natural Science and Technology, Okayama University, Tsushima, Kita-ku, Okayama 700-8530, Japan
[e] Advanced Nano-Characterization Center, National Institute for Materials Science, Sakura, Tsukuba 305-0003, Japan
[f] Graduate School of Life Science, Hokkaido University, Kita-ku, Sapporo 060-0812, Japan

ABSTRACT

Novel Ca-containing tubular silica nano-fibril (NT) mats were fabricated using reassembled collagen fibrils as the template, and their structure, *in vitro* apatite formation, and cell compatibility were studied. The reassembled collagen fibrils were soaked in a Stöber type sol-gel precursor mixture of tetraethoxysilane/ethanol/water/ammonium hydroxide/$CaCl_2$ for 24 h at room temperature to produce silica-coated collagen fibril hybrids, which were subsequently calcined at 600 °C to produce Ca-containing silica NT mats. When soaked in Kokubo's simulated body fluid, the resultant mats deposited petal-like apatite crystallites: the silica NTs had in vitro bioactivity. They favored attachment and growth of osteoblast-like MC3T3-E1 cells, and the addition of Ca(II) was found a key factor to stimulate cell proliferation and differentiation.

1. INTRODUCTION

Several silica-based glasses and glass-ceramics have been widely applied to clinics as they promote bone tissue generation and form strong bond with the surrounding tissues. It is widely accepted that Si-OH groups serve as active sites for nucleation and growth of apatite under the body conditions, and that the Ca ions released from those materials promote apatite deposition with increasing the super-saturation of blood plasma for apatite precipitation.[1] Similar effects have been observed for silica nano- and micro-particles[2-3] in the Stöber-type[4] systems. In addition, since the experiments by Carlisle,[5] Si(IV) is long known as one of the essential elements for chicken skeleton growth, and Xynos et al.[6] pointed out that Si(IV) from some silicate glasses stimulated genes relevant to bone cell proliferation. As nano-or micro-fibrils have a larger surface area, and so many active sites may be available, silica in unwoven mats or sheets will be good candidates for biologically active devices. Even better effects are expected when the fibrils are hollow, or tubular.

Tubular inorganic structures of TiO_2 and carbon, sometimes exhibit biologically useful behaviors, such as biocompatibility and bioactivity or spontaneous apatite deposition under the body conditions. For example, Balasundaram et al. electrochemically grew TiO_2 tubular arrays on metallic Ti and immobilized morphogenetic protein-2 to increase osteoblast adhesion.[7] After Tuschiya et al.[8], similar anodic TiO_2 tubes deposited apatite in the Kokubo's simulated body fluid (SBF). Carbon NTs when coated on highly porous bioglass scaffolds were known to induce apatite deposition and to promote bone tissue generation.[9-10] When those tubes were in mats or sheets, they would be fitted to defects of any shape. Hayakawa et al. prepared polycrystalline rutile nano-tubes with cotton fibrils as the template, and showed their *in vitro* bioactivity.[11] Thus, this suggests that fibrous templates lead to tubular ceramic fibrils.

Therefore, in the present study, we fabricated novel Ca-containing silica NT mats using collagen fibrils as the template, and investigated their in vitro apatite forming ability. A preliminary study was carried out on their biocompatibility due to culturing osteoblast-like MC3T3-E1 cells.

2. EXPERIMENTAL

Fig. 1 illustrates the procedure of fabricating the silica NT mats. Briefly, type I porcine collagen solution in dilute HCl was neutralized at 37 °C for 3 h with phosphate buffer solution to yield reassembled porcine collagen fibrils. They were subsequently soaked in a Stöber-type sol-gel precursor mixture,[2-3] ethanol (30 mL), water (5 mL), ammonium hydroxide (0.3 mL, 25%), tetraethoxysilane (TEOS; 5 mL), calcium chloride (0 or 3 mmol), for 24 h at room temperature. The silica-coated collagen fibrils were finally calcined at 600 °C for 2 h to produce the silica NT unwoven mats. Those from the precursor mixture with and without calcium chloride were coded as Si and SiCa, respectively.

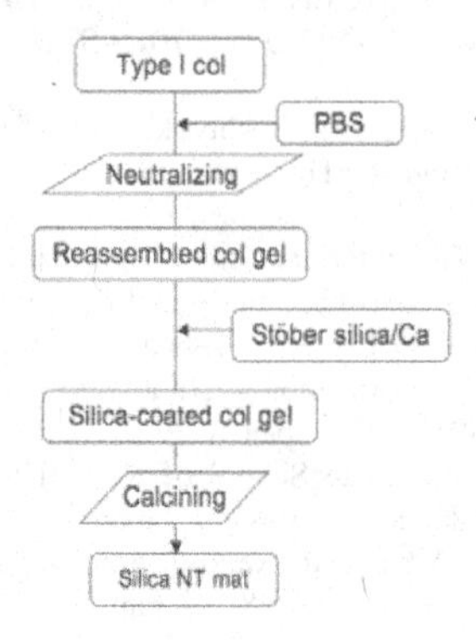

Fig. 1. Procedure of the silica nano-tube fibril mats with collagen fibrils as the templates.

In vitro bioactivity of the resultant unwoven mats was examined to detect spontaneous apatite deposition when soaked in Kokubo's simulated body fluid (SBF). Their microstructure and morphology were observed under a scanning electron microscope (SEM; JSM-5600LV, JEOL, Tokyo, Japan) and transmission electron microscope (TEM; JEM-2100F, JEOL, Tokyo, Japan). The biological response was evaluated by culturing them with osteoblast-like MC3T3-E1 cells. Morphology and size of the cells on the silica NT mats were observed under a field emission scanning electron microscope (FE-SEM; JSM-6500F, JEOL, Japan).

3. RESULTS

The SEM images in Fig. 2 show the microstructure of silica NT mats Si (a) and SiCa (b). Both samples consisted of fibrils with different diameters. The average fibril diameter for Si and SiCa was 150 ± 10 nm and 190 ± 20 nm, respectively. The addition of $CaCl_2$ to the precursor mixture increased the fibril diameter by ~ 40 nm.

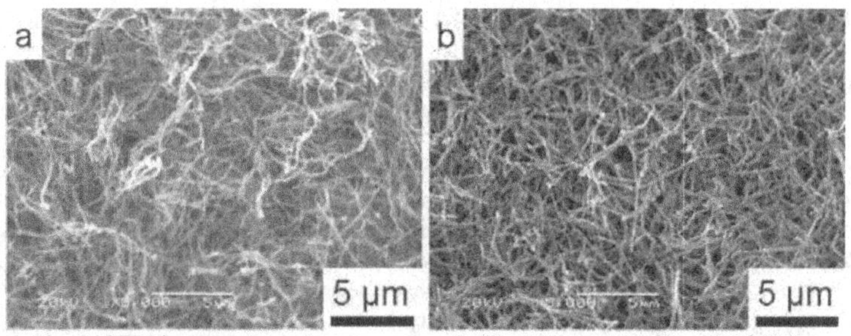

Fig. 2. SEM images: the microstructure of silica NT mats (a) Si and (b) SiCa.

Fig. 3 shows TEM images of the fibrils in the silica NT mats (a) Si and (b) SiCa. Clear contrast between the center and periphery parts was found for both samples, indicating the formation of silica NTs. Both samples have a similar inner diameter of ~ 120 nm. However, the shell thickness of them was different: ~ 30 nm for Si and ~ 50 nm for SiCa. In contrast, the addition of Ca ions led to rougher and thicker silica shells. It is consistent with our previous study.[2]

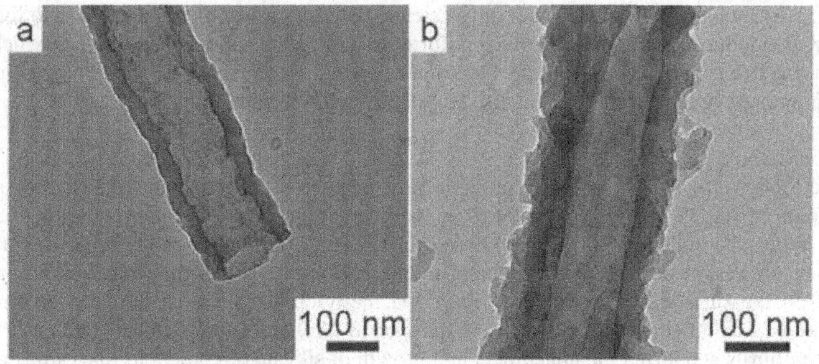

Fig. 3. TEM image of one of the fibrils of the silica NT mats (a) Si and (b) SiCa, indicating the fibrils are hollow.

Fig. 4 shows the SEM images of silica NT mats (a) Si and (b) SiCa after soaked in SBF for one week. Comparing the microstructures of sample Si in Fig. 4(a) and Fig. 2(a), no significant change in morphology was found, regardless of soaking in SBF. The morphology of sample SiCa in Fig. 4(b) is different from that in Fig. 4(a): new petal-like substances were deposited on the surface when soaked in SBF. Those new substances were apatite after the X-ray diffraction profile for sample SiCa soaked in SBF and the silica NTs provided the active sites for apatite deposition. Moreover, the addition of Ca ions promoted apatite deposition.

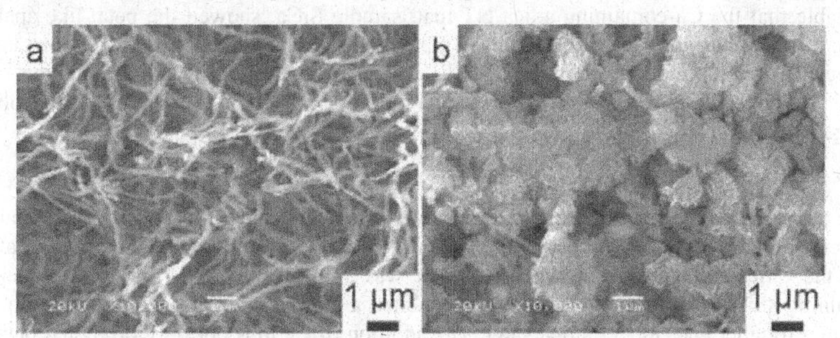

Fig. 4. SEM images of silica NT mats (a) Si and (b) SiCa, after soaked in SBF for 7 d.

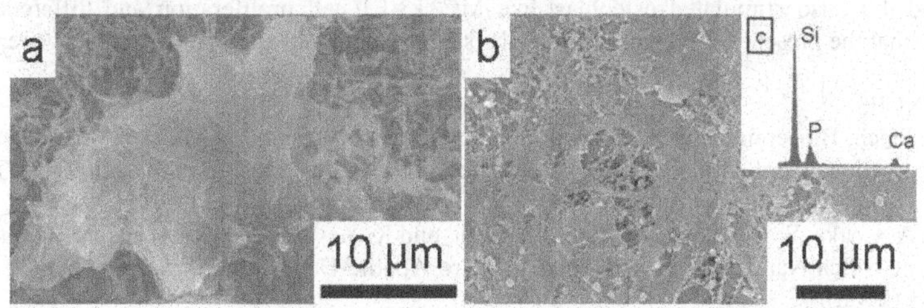

Fig. 5 SEM images of the silica NT mats (a) Si and (b) SiCa as well as (c) an EDS spectrum (insert) of SiCa after both mats were cultured with osteoblast-like MC3T3-E1 cells for 21 d.

Fig. 5 shows the SEM images of the morphology of osteoblast-like MC3T3-E1 cells cultured on silica NT mats (a) Si and (b) SiCa for 21 d. The cells in Fig. 5(a) extended and grew on the surface of the silica NT sample Si, without Ca ions, indicating that the Ca-free silica NT mat supported cell attachment and growth. Even stimulated cell growth was found for silica NT sample

SiCa: Fig. 5(b) shows that those cells not only extended on the surface of the silica NTs to form cell layers, but also induced formation of collagen fibrils as well as the spherical substances which were rich in Ca and P, as confirmed in an EDS spectrum (insert). Therefore, the Ca-containing silica NTs not only promoted cell proliferation, but also stimulated cell differentiation.

4. DISCUSSION

In the Stöber-type sol-gel system, ammonia was the base catalyst and initiated hydrolysis and condensation of TEOS monomers to produce silica colloid particles of several nm to μm in diameter. When the collagen fibrils were soaked in such a sol-gel colloidal system, two probable mechanisms are responsible for the formation of silica NTs. One is that the silica colloid particles deposited on the surface of the collagen fibrils via the hydrogen bonding between proteins and silanol groups. The other is that the active groups such as NH_2 and COOH groups on the surface of the collagen fibrils supported the attachment and growth of silicate oligomers to form silica particle layers via hydrogen bonding or electrostatic interaction. Ono et al.[12] fabricated silicate-collagen fibril hybrids from a neutral mixture of collagen fibrils and TEOS while Eglin et al.[13] obtained those from a neutral mixture of collagen gel and tetramethoxysilane. They proposed that electrostatic interaction and hydrogen bonding worked among anionic silica oligomers and the cationic collagen fibrils to derive the silica-collagen hybrids. In the present case, ammonia was used as catalyst to promote the formation of silica particles in the solution. Therefore, both mechanisms work together in a harmonized way.

Although it is widely accepted that the Si-OH groups in the hydrated silica gel layer provide heterogeneous nucleation sites for apatite deposition,[1, 14-15] Tsuru et al. pointed out that the released Ca(II) from their organic-inorganic hybrids predominantly worked for apatite deposition.[16] Thus, it is reasonable that the Ca-containing silica NT mat, sample SiCa, showed the petal-like apatite in Fig. 4(b), or had *in vitro* bioactivity. Moreover, Ren et al. reported the addition of Ca ions into the silicate-gelatin hybrids indeed promoted MC3T3-E1 cell proliferation and differentiation,[17] with which the present study agreed well. Therefore, the present Ca-containing silica NTs are applicable to bone generation.

5. CONCLUSION

The reassembled collagen fibrils were coated with silica sol particles after they were soaked in the Stöber-type sol-gel system of tetraethoxysilane/ethanol/water/ammonium hydroxide/$CaCl_2$ to produce the silica-collagen fibril hybrids, which was further calined at 600 °C to yield the Ca-containing silica NT mats. The silica NTs exhibited a tubular structure and the addition of Ca ions led to a rougher and thicker silica shell. The addition of Ca ions not only promoted petal-like apatite deposition when those Ca-containing silica NTs were soaked in the Kokubo's simulated body fluid, but also stimulated osteoblast-like MC3T3-E1 cell proliferation and differentiation, indicating that the present Ca-containing silica NTs have a potential application to bone generation.

REFERENCES

[1]a) L. L. Hench, Bioceramics from Concept to Clinic, *J. Am. Ceram. Soc.*, **74**, 1487-510 (1991); b) T. Kokubo, Bioactive Glass Ceramics: Properties and Applications, *Biomaterials*, **12**, 155-63 (1991).

[2]S. Chen, A. Osaka, S. Hayakawa, K Tsuru, E. Fujii, and K. Kawabata, Microstructure Evolution in Stöber-Type Silica Nanoparticles and Their In Vitro Apatite Deposition, *J. Sol-Gel Sci. Technol.*, **48**, 322-35 (2008).

[3]S. Chen, S. Hayakawa, E. Fujii, K. Kawabata, K Tsuru, and A. Osaka, Sol–Gel Synthesis and Microstructure Analysis of Amino-Modified Hybrid Silica Nanoparticles from Aminopropyltriethoxysilane and Tetraethoxysilane, *J. Am. Ceram. Soc.*, **92**, 2074-82 (2009).

[4]W. Stöber, and A. Fink, Controlled Growth of Mono-disperse Silica Spheres in the Micron Size Range, *J. Colloid Interface Sci.*, **26**, 62-69 (1968).

[5]E. M. Carlisle, Silicon: an Essential Element for the Chick, *Science*, **178**, 619-21 (1972).

[6]I. D. Xynos, A. J. Edgar, L. D. K. Buttery, L. L. Hench, and J. M. Polak, Gene-Expression Profiling of Human Osteoblasts Following Treatment with the Ionic Products of Bioglass 45S5 Dissolution, *J. Biomed. Mater. Res.*, **55**, 151-7 (2001).
[7]G. Balasundaram, C. Yao, and T. J. Webster, TiO_2 Nanotubes Functionalized with Regions of Bone Morphogenetic Protein-2 Increases Osteoblast Adhesion, *J. Biomed. Mater. Res. A*, **84**, 447-53 (2008).
[8]H. Tsuchiya, J. M. Macak, L. Müller, J. Kunze, F. Müller, P. Greil, S. Virtanen, and P. Schmuki, Hydroxyapatite Growth on Anodic TiO_2 Nanotubes. *J. Biomed. Mater. Res., Part A*, **77**, 534-40 (2006).
[9]T. Akasaka, F. Watari, Y. Sato, and K. Tohji, Apatite Formation on Carbon Nanotubes, *Mat. Sci. and Eng. C*, **26**, 675-8 (2006).
[10]R. Boccaccini, F. Chicatun, J. Cho, O. Bretcanu, J. A. Roether, S. Novak, and Q. Z. Chen, Carbon Nanotube Coatings on Bioglass-Based Tissue Engineering, *Adv. Funct. Mater.*, **17**, 2815-22 (2007).
[11]S. Hayakawa, J.-F. Liu, K. Tsuru, and A. Osaka, Wet Deposition of Titania-Apatite Composite in Cotton Fibrils, J. Sol-Gel Sci. Technol., **40**, 253-258 (2006).
[12]Y. Ono, Y. Kanekiyo, K. Inoue, J. Hojo, M. Nango, and S. Shinkai, Preparation of Novel Hollow Fiber Silica Using Collagen Fibers as a Template, *Chem. Lett.*, **6**, 475-6 (1999).
[13]D. Eglin, G. Mosser, M. M. Giraud-Guille, J. Livage, and T. Coradin, Type I Collagen, a Versatile Liquid Crystal Biological Template for Silica Structuration from nano to microscopic Scales, *Soft Matter.,* **1**, 129-31, (2005).
[14]H. Takadama, H.-M. Kim, T. Kokubo, and T. Nakamura, Mechanism of Biomineralization of Apatite on a Sodium Silicate Glass: TEM–EDX Study In Vitro, *Chem. Mater.*, **13**, 1108-13 (2001).
[15]T. Kokubo, H. Kushitani, C. Ohtsuki, S. Sakka, and T. Yamamuro, Effects of Ions Dissolved from Bioactive Glass-Ceramic on Surface Apatite Formation, *J. Mater. Sci.-Mater. Med.*, **4**, 1-4 (1993).
[16]K. Tsuru, Y. Aburatani, T. Yabuta, S. Hayakawa, C. Ohtsuki, and A. Osaka, Synthesis and in vitro Behavior of Organically Modified Silicate Containing Ca Ions, *J. Sol–Gel Sci. Technol.*, **21**, 89-96 (2001).
[17]L. Ren, K. Tsuru, S. Hayakawa, and A. Osaka, Novel Approach to Fabricate Porous Gelatin–Siloxane Hybrids for Bone Tissue Engineering, *Biomaterials*, **23**, 4765-73 (2002).

TISSUE INGROWTH IN RESORBABLE POROUS TISSUE SCAFFOLDS

Janet Krevolin[1], James J. Liu[1], Adam Wallen[1], Kitu Patel[1], Rachel Dahl[1], Hu-Ping Hsu[2], Cathal Kearney[2], Myron Spector[2]

[1]Bio2 Technologies
12-R Cabot Road
Woburn, MA 01801

[2]VA Boston Healthcare System
Tissue Engineering Lab
150 S. Huntington Ave
Boston, MA 02130

ABSTRACT

Highly porous scaffolds fabricated using a proprietary Cross-Linked Microstructure (CLM) process in a resorbable (bioactive) glass composition were subjected to an *in vivo* animal study to determine and evaluate the feasibility for use as a weight-bearing tissue engineering scaffold. The CLM scaffolds were manufactured by processing resorbable glass fibers to create a cross-linked microstructure of an interconnected open pore network. The fabrication of such novel scaffolds involves pore formation and strengthening sintering techniques which lead to a well defined cross-linked microstructure with desired mechanical properties. A highly interconnected pore network is created by selection and arrangement of pore-formers and fibers. The porosity can be controlled from 20 to 75% with a pore size range from 50 to 750μm.

The CLM structure in a resorbable composition with open and interconnected porosity exhibits the potential to provide adequate mechanical properties and bioactivity at the same time. *In vivo* results suggest that new bone has formed in the pores and around the fibers of the CLM implants. There is also evidence of bone formation directly on the surface of the CLM implants, with continuity of the mineral phase between the bone and implant demonstrating direct bonding to the scaffold.

An optimized porous CLM resorbable glass scaffold, with a bonded fiber structure and interconnected porosity is promising for applications in orthopedic tissue engineering as it combines high strength and bioactive properties. The potential applications of this high strength, resorbable glass fiber-based scaffold include load-bearing implants for treating problems of the spine and extremities.

INTRODUCTION

Prosthetic devices and other implants are often required for the replacement, augmentation, or repair of diseased or deteriorated musculoskeletal structures. In some cases the goal is to accelerate and enhance the body's own reparative processes after musculoskeletal injuries resulting from severe trauma or degenerative disease. Various types of synthetic implants have been developed for tissue engineering applications in an attempt to provide a device that mimics the properties of natural bone tissue and promotes healing and repair of tissue.[1-4]

The challenge in developing a resorbable tissue scaffold using biologically active and resorbable materials is to attain load-bearing strength with porosity sufficient to promote ingrowth of bone tissue. Conventional bioactive bio-glass and resorbable bio-ceramic materials in a porous form are not known to be inherently strong enough to provide load-bearing strength as a synthetic prosthesis or implant.[2,4] Similarly, conventional bioactive materials in a form with sufficient strength do not exhibit a pore structure with sufficient porosity to accommodate the amount of bone ingrowth that may be required for substantive fixation of the device.

Fiber-based structures are generally known to provide inherently higher strength to weight ratios, given that the strength of an individual fiber can be significantly greater than powder-based or particle-based materials of the same composition. A fiber can be produced with relatively few discontinuities that contribute to the formation of stress concentrations for failure propagation. By contrast, a powder-based or particle-based material requires the formation of bonds between each of the adjoining particles, with each bond interface potentially creating a stress concentration.

Furthermore, a fiber-based structure provides for stress relief and thus, greater strength, because when the fiber-based structure is subjected to strain the failure of any one individual fiber does not propagate through adjacent fibers. Accordingly, a fiber-based structure exhibits superior mechanical strength properties over a powder-based structure of the same composition with an equivalent size and porosity.[7-8]

In this work we demonstrate a family of novel, microstructurally-ordered materials – the fibrous cross-linked microstructure (CLM) ceramic materials. Bio-ceramics produced with the fibrous CLM ceramic materials possess the advantages of a high porosity macrostructure and the high performance mechanical properties of the fibrous microstructure. There are significant differences between fibrous CLM materials and fibrous materials, such as those with sintered fiber structures, documented in previous literature.[9-16] The aim of this study was to characterize the fibrous CLM bio-ceramics and demonstrate their performance advantages for bone tissue engineering applications.

EXPERIMENTAL PROCEDURES

Fabrication

A CLM technique was utilized to produce test samples with different porosities and pore interconnectivity. The CLM process was used to produce scaffolds with a range of porosities, pore sizes, mechanical properties and *in vitro* degradation rates. The diameter and composition of the fibers, along with the size and composition of the chosen pore-formers and binders, influence the resulting pore size and pore size distribution of the structure. The fibrous CLM process includes a fibrous and powder mixture extrusion procedure. Fibrous CLM bio-ceramics were extruded from ceramic paste which was prepared by mixing >50 wt% of fiber with addition of binders and pore formers. In an example of a CLM bio-glass material 13-93 glass fiber was used as a raw material and mixed with pore-formers and sintering aids. 13-93 glass, with the following composition: 6mol% Na_2O; 7.9% K2O; 7.7% MgO; 22.1% CaO; 0% B2O3; 1.7% P2O5; and 54.6% SiO2, was drawn using in-house fiber drawing equipment. Both process yield and fiber consistency were evaluated as the fiber composition, diameter and process parameters were altered. Starting fibers were chopped to size with lengths ranging from 0.5 to 2 cm, and diameters greater than 15 μm. These constituents were mixed in a high-shear sigma-blade mixer for ~60 minutes until a ceramic paste was formed. The pastes were then de-aired, formed into a billet and extruded into a 12.5 mm round rods using a stainless die in a hydraulic ram extruder.

Mechanical Analysis

The compressive strength, elastic modulus under compression, and respective failure modes were determined, since those properties are critically important in several orthopedic bone scaffolding applications. A standard test protocol (e.g. ASTM F451-08) was modified to fit the bioactive glass scaffold testing. The axial compressive strength was measured from cored samples on a Statec universal testing machine under load control. Compressive strength was measured in the direction parallel to the axis of extrusion. The microstructure was analyzed using scanning electron microscopy (SEM) performed at Alfred University. The operating voltage used was 20 kV.

Osteconductive Analysis

In order to test the osteoconductive properties of the fabricated scaffolds both *in vitro* and *in vivo* testing were carried out to confirm that a calcium phosphate layer forms on the surface, which can then precipitate to hydroxyapatite (HAp).

In vitro testing

For the *in vitro* testing, porous bodies were immersed in simulated body fluid (SBF) to evaluate their degradation process. The composition of the immersion fluid and the method to produce it were based on the work of Kokubo et al.[17,18] Previous studies have shown that immersion temperature (T), time (t) and surface-area-to-volume ratio (SA/V) are parameters which affect the *in vitro* degradation process of bioactive materials.[17-19]

Samples of the raw material fiber and scaffolds were placed in polystyrene bottles containing SBF with ion concentrations nearly equal to human blood plasma.[17,18] The bottles with the samples and SBF were maintained at 36.5 °C in a shaking water bath for 1, 7, and 14 days respectively without refreshing the soaking medium. The sample surface area to SBF volume (SA/V) ratio of 0.1 cm^{-1} was used for all the test samples. After various soaking periods, the samples were filtrated and gently rinsed twice with ethanol to remove SBF followed by drying in vacuum at 80 °C. Three sets of analysis were performed for samples with *in vitro* conditioning: (1) changes in pH of the solutions; (2) change in mass; and (3) SEM and compositional analysis of the glass surface after immersion in SBF. The HAp which formed on the surface of the fibers and scaffolds was characterized by x-ray diffraction (XRD) and SEM.

In vivo testing

Pilot *in vivo* rabbit studies were performed, with harvest points at 2, 6 and 8 weeks. New Zealand white rabbits 6 months in age were used, with two rabbits at each harvest point. A rod approximately 8 mm in length and 4 mm in diameter of bio-glass CLM was implanted into the femoral condyle region of both legs of each rabbit in order for the implant to reside mainly in cancellous bone. The rod was inserted using an interference fit by first drilling a 4-mm diameter hole in the femur and then press-fitting the implant into the hole. The rabbits were injected with calcein fluorochrome labels twice, with the second label given 4 days prior to sacrifice. The rabbits received pain medication and were monitored. The protocol was approved by the Institutional Animal Care and Use Committee of the laboratory facilities conducting the animal studies.

Post-sacrifice, the femoral condyles were dissected and were subject to microCT (eXplore, GE) scanning. Undecalcified samples of the implants and surrounding bone were embedded in a polymer medium and cut with a precision diamond wafer blade. The cut surfaces through the middle of the implants were polished and examined in a SEM in the backscattered electron imaging (BEI) mode to determine if bone formed directly on the CLM surface and to confirm the bone ingrowth into the pores (as revealed by microCT). For histology, sections through the implant and bone (~100μm thick) were cut with a precision saw, glued to a glass slide, ground and polished to about 40μm thick. The sections were then stained (e.g. toluidine blue) and examined in a light microscope to further confirm bone bonding to the CLM and bone growth into the pores.

RESULTS AND DISCUSSION

The CLM microstructure produces a combination of high strength at high porosity and high material permeability for a given pore size distribution. The unique material properties of CLM scaffolds are due to the cross-links between fibers and the crystal and grain structure induced through chemistry and sintering. Fibers are arranged in a 3-D interconnected matrix during mixing and extrusion. Fibers are cross-linked through sintering and *in situ* chemical reactions. The sintering process is necessary to obtain a mechanically rigid three-dimensional CLM substrate. The strength of the substrate is derived from the fibers and bonds between overlapping and adjoining fibers within the structure.

Porosity and pore size distribution are important parameters for the scaffold application, with high porosity in scaffolds allowing for cellular distribution.[12] Porosity can, however, adversely affect important mechanical characteristics of a scaffold, for instance it is difficult to achieve a high porosity with high strength. This challenge may be overcome by re-engineering a porous scaffold using fibrous structure. In so doing, it is possible to extend the porosity range at the same time, while maintaining adequate mechanical strength. High porosity is one of the key characteristics of fibrous structured materials. The porosity of a fibrous structure is created by open space between fibers. A highly porous fibrous structure is also utilized in many other applications, such as fiber insulation and fiber filtration. The results indicate that extruded fibrous CLM ceramics have porosities greater than 50% and can reach up to 80% porosity and maintain equivalent strength. The porosity of extruded fibrous CLM ceramics depends on several factors, such as fiber diameter, pore-former, chemical reactions and sintering.

The CLM pore structure is different from traditional particle based or fiber based ceramics. Pore structure of sintered particle or fibers is limited by sintering conditions, such as temperature and time. Pore structure of CLM ceramics can be controlled by chemistry and sintering. CLM in a 13-93 bio-glass composition exhibits a bimodal pore size distribution with the large pore size range from 50 to 750 μm. This is particularly important for applications that require load bearing, where large pores or open channels are needed for vascularization.

SEM of CLM scaffolds are shown in Figure 1. These images indicate that the cross-linked microstructure ranged from fibrous bio-active glass fiber to fine crystalline fiber. The cross-linked microstructure can be provided in an extruded textured pattern that is formed from ceramic fibers. The resulting cross-linked microstructure has a greater number of connection points for additional crossed-linked bonding.

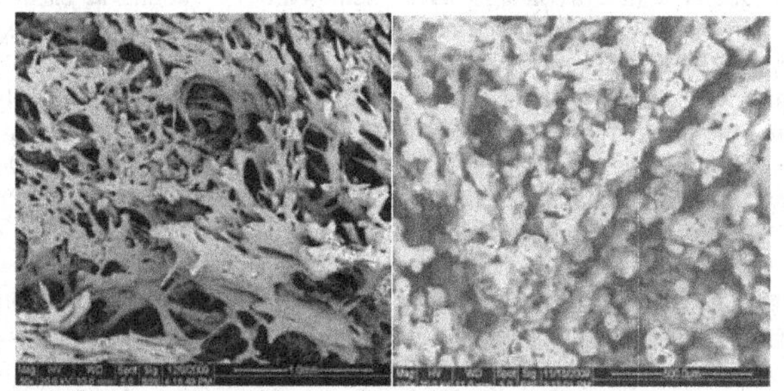

Figure 1. SEM of CLM Bio-glass (Back Scattered Electron Image)

The process of bone graft incorporation is similar to the bone healing process that occurs in fractured long bones.[6] Bone grafts are also strongly influenced by local mechanical forces during the remodeling phase. Mechanical demands modulate the density, geometry, thickness and trabecular orientation of bone, allowing optimization of the structural strength of the graft. The fibrous CLM has much improved mechanical properties compared to particle-based microstructures, because of the inherent characteristic of fiber. Fibrous CLM materials have adequate mechanical strength at high porosity (Figure 2). There is a correlation between strength and porosity for CLM, and this correlation may be different from porous particle-based ceramics. Accurate prediction of the strength of porous CLM materials requires more adequate information for comparison. It is necessary to point out that the differential results of theoretical predication models may arise from irregular shapes, varied size and random distribution of the pores and others.

For similar porosities, CLM bio-ceramics have higher mechanical properties compared to sintered fiber-based bio-ceramics. The improved mechanical performance of CLM ceramics may be due to the unique microstructure and chemistry. The microstructural evolution of CLM ceramic materials with increasing porosity is related to 3-D connectivity. Connectivity represents the contact area between fibers or fibrous grains. The increasing connectivity results in increased mechanical strength, however, the bonding strength between the fibers is important for increased load bearing (Figure 2).The minimum contact solid area (load-bearing area) which is the actual sintered or the bond area is significantly different between particle-based and fiber-based materials.

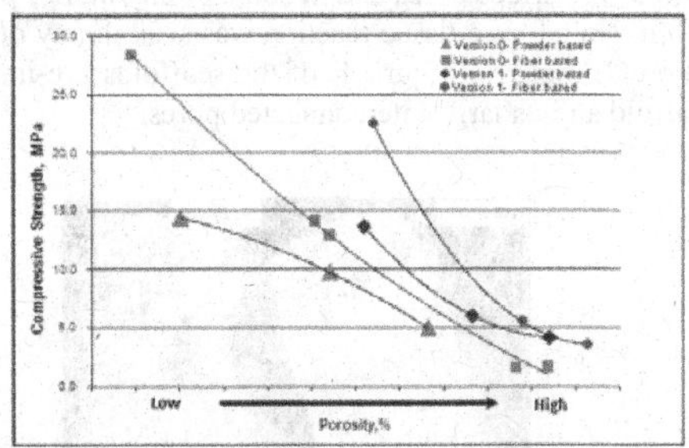

Figure 2. Relationship between porosity and compressive strength in powder and fibrous-based scaffolds

Based on the promising CLM microstructures and the unique elastic plastic mechanical response, an *in-vitro* study was undertaken to evaluate the response of bioactive glass scaffolds prepared by the CLM extrusion process in SBF. Bioactive glass with the 13-93 composition was used because of its proven bioactivity,[10] as well as its ability to support cell proliferation and function.[11,12] The glass is also approved for *in-vivo* use in the United States and elsewhere.

The results indicated that similar to the 13-93 powder,[16] fibers could also develop bone-like HAp layers on its surface when exposed to SBF. It is a common notion that bone-like HAp, which is a calcium phosphate that resembles bone mineral, plays an essential role in the formation, growth and maintenance of the tissue-biomaterial interface.[19-22] Therefore, the present study suggests that scaffolds made from CLM 13-93 bio-glass are bioactive. Figure 3 shows SEM images of 13-93 CLM scaffold after soaking in SBF for 14 days. From Figure 3a, it can be seen that bone-like HAp forms on the surface. The energy dispersive x-ray (EDAX) spectra, shown in Figure 3c, demonstrates that the characteristic peaks of SiO_2 disappeared as HAp formed. After prolonged soaking for 7 and 14 days, the crystalline identity of HAp was detected in the XRD patterns.

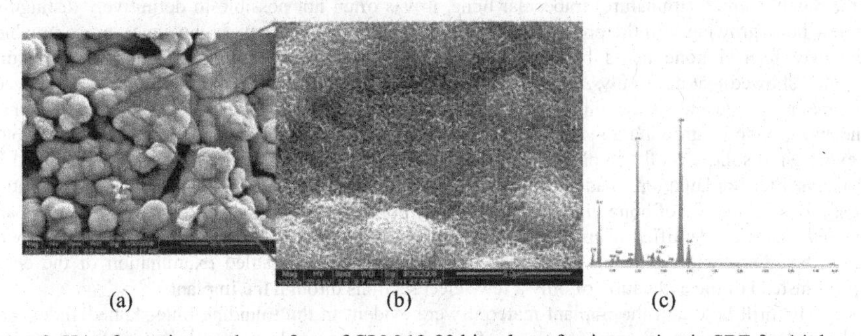

Figure 3. HAp formation on the surface of CLM 13-93 bio-glass after immersion in SBF for 14 days. (a) SEM of bone-like HAp on the surface of the samples. (b) Higher magnification SEM revealed the nano-crystals making up the HAp layer. (C) EDAX spectra indicating new HAp and disappearance of SiO2 characteristic peaks

The scaffolds implanted in the rabbits had a porosity of 60±10%, a pore size distribution from 50-750 µm, an average pore diameter of 375 µm , and a compressive strength ranging from 6 and 23 MPa. Both microCT and histomorphometry demonstrated bone formation between the implant surface and the host bone as well as in surface pores and in some of the interior pores of the implant, 2, 6, and 8 weeks post implantation. No adverse tissue reaction was seen in any of the sections examined at the various time points. MicroCT images (Figure 4) of the scaffolds pre-implantation showed the highly porous nature of the scaffold and its large interconnected pores.

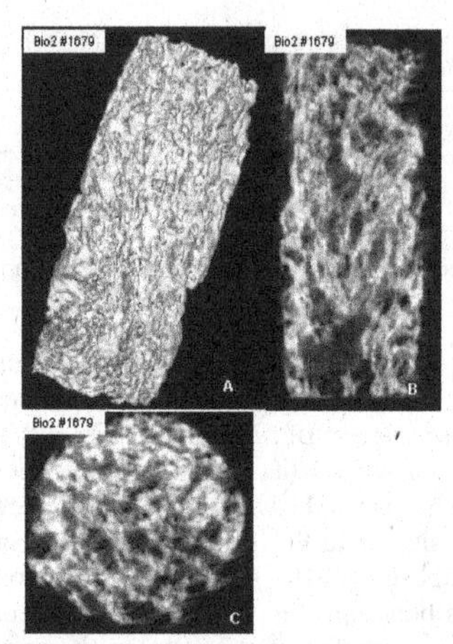

Figure 4. MicroCT images prior to implantation. The scaffolds were approximately 4mm in diameter. (A) Reconstructed images of the surface (B) "Y-slice" image through the middle of the sample (C) "Z-slice" image through the sample

The radiodensity of the CLM implant (x arrows in Figure 5a and b) was slightly lower than that of the surrounding mature trabecular bone. Because the radiodensity of the implant was similar to that of the newly formed (immature) trabecular bone, it was often not possible to definitively distinguish the new bone growing into the pores from the implant material. There were, however, pores in which the newly formed bone could be distinguished from the implant material by its microstructure. MicroCT showed that after only 2 weeks of implantation, trabecular bone formed in the gaps between the implant surface and surrounding bone and in surface pores (y arrows in Figure 5a and b) This new bone appeared to be growing up to the implant surface (*i.e.*, "osseointegration"). This new bone would be expected to solidly fix the implant to bone. In fact, it is likely that enough new bone had formed by 7-10 days after implantation to adequately stabilize the implants. Also of note was the clear indication in the microCT images of bone ingrowth into interior pores of the implants (z arrows in Figure 5a and b). These features identified in microCT were confirmed by the undecalcified SEM/BEI and ground section histology. The value of the microCT analysis is that it enabled examination of the entire implant instead of the evaluation of only a few select sections through the implant.

The drill hole and the implant material were evident in the toluidine blue-stained slides, at 2 weeks and at 6 weeks (Figure 6). Newly formed bone appeared as fine spicules (blue staining) interlaced and webbed around the implant fibers. The newly formed bone occupied a large portion of the implant area.

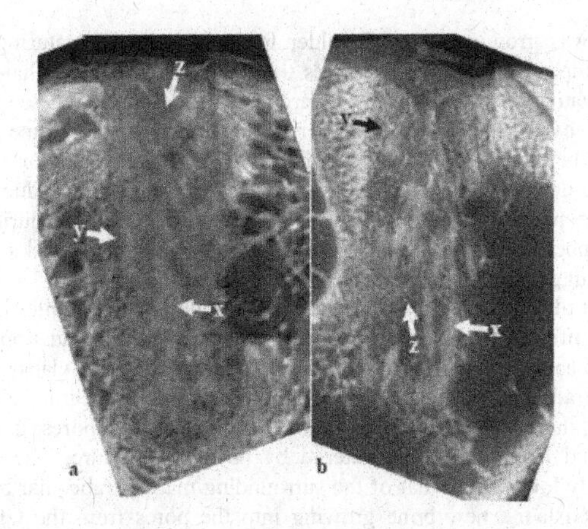

Figure 5. MicroCT images after 2 weeks in-vivo. The CLM implant is shown by the x arrows. Examples of newly formed bone are shown; y arrows show the gaps between the implant perimeter and surrounding cancellous bone and in surface pores, z arrows show the implant

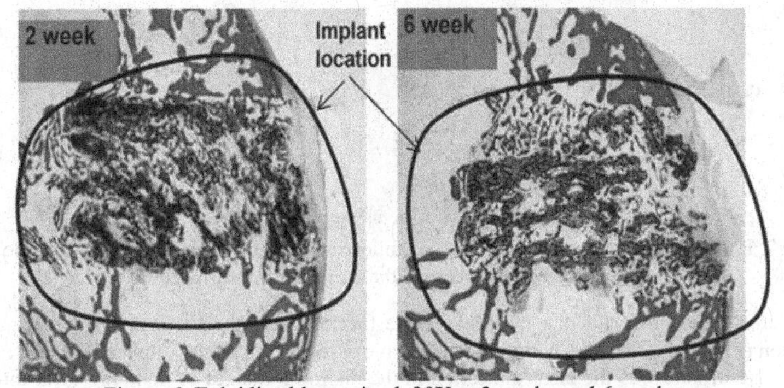

Figure 6. Toluidine blue stained, 20X at 2 weeks and 6 weeks

We observed in the numerical microCT data that throughout the implantation period the bone volume and thickness of trabecular structure increased and the open space between the trabecular bone structures decreased. Histological slides of the specimens exhibited newly formed bone incorporating the implant. The percentage of the implantation region containing newly formed bone increased more than double from week 2 to 6, reflecting ongoing bone ingrowth.

The newly formed bone comprised regions of mixed woven and lamellar bone. By week 6, large regions of lamellar bone were seen in the implantation site. The amount of fluorochrome labeled bone in the defect region was similar between weeks 2 and 6, indicating similar degrees of bone mineralization. An increase in bone volume but a similar level of bone mineralization indicated that new woven bone was growing in, while older formed bone was maturing into lamellar bone. The structural model index (SMI) which estimates the trabecular plate-rod characteristics of the material, becomes more negative as the structure undergoes remodeling.[26] We see the SMI value becoming increasingly negative over the implantation period which may be due to the plate-like structures in the scaffold at week 0 being resorbed and becoming more rod like. Based on the SMI trend, we believe that the connective density increasing between week 0 and 2 is likely due to trabecular rods being broken in multiple pieces and the fenestration of trabecular plates during remodeling,[27] thereby increasing the number of connections. We observed that between weeks 2 and 6 the connectivity density decreases suggesting that new tissue is maturing.

The amount of residual implant material was similar in the samples harvested at week 2 and 6, indicating a slow initial degradation rate. A slow initial rate of degradation would be desirable for a bioresorable, load bearing implant, as it would allow enough time to elapse for mature bone ingrowth to occur before degradation caused a decrease in the implants load bearing capacity.

At 8 weeks, there is evidence of new bone formation in the pores, in which newly formed bone can be distinguished from the CLM material by its microstructure. The radiodensity of the CLM material was slightly lower, than that of the surrounding mature trabecular bone, making it difficult to definitively distinguish the new bone growing into the pores from the CLM material. Further, the macroporous structure of the CLM material was no longer apparent on the implanted microCT images at 8 weeks, which is believed to be due to the filling of the pores. It is difficult to fully interpret whether the CLM material remains at week 8 from the microCT images alone; however, the top of the scaffold was seen at the cortical bone surface on explant, suggesting that some of the scaffold remains (Figure 7).

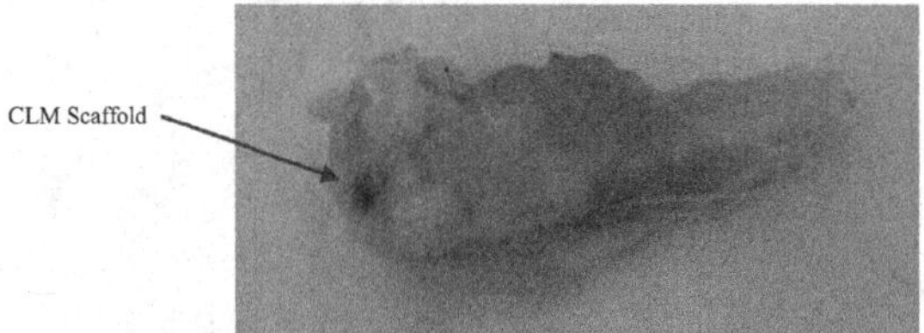

Figure 7. Photograph of the rabbit distal femur following explant at week 8. The top of the CLM scaffold can be seen on the cortical bone surface

MicroCT demonstrated that trabecular bone formed up to the scaffold surface by week 8 of implantation (Figure 8). The new bone appears to have osseointegrated the implant, fixing it in place in the bone. Also of note was the clear indication in the microCT images of bone ingrowth into interior pores of the implants (Y arrows in Figure 8). As shown by BEI in Figure 9, by 8 weeks new bone had been formed around and within the pores of the CLM scaffold. There is clear evidence of bone formation on the surface of the scaffolds, as well as continuity of mineral between the implant and newly forming bone. There appeared to be less remaining bio-glass in the implant sites at 8 weeks compared to the 2-week findings. The struts/members of trabecular bone in the implanted sites appeared to be thicker than seen at 2 weeks, indicating maturation of the trabecular bone.

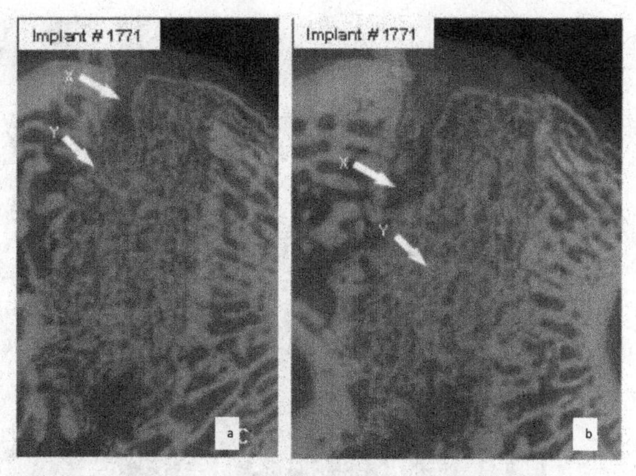

Figure 8. MicroCT cross-section of the CLM scaffold in the rabbit distal femur at 8 weeks. X arrows point to gaps between the scaffold and bone (due to surface irregularities); Y arrows point to regions demonstrative of new bone formation within the scaffold.

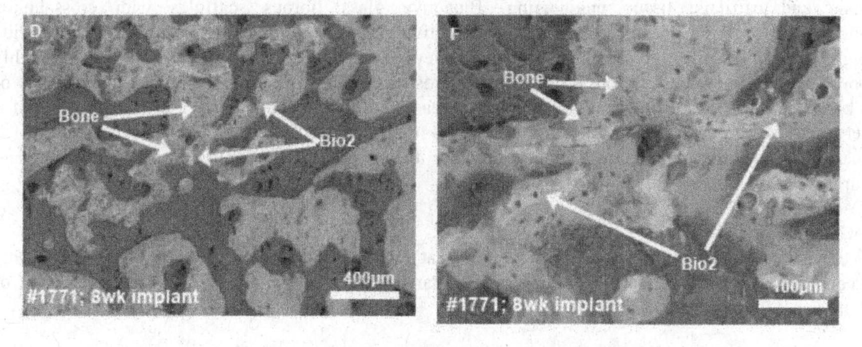

Figure 9. BEI micrographs following 8 weeks of implantation

EDX was used to interrogate the mineral distribution and scaffold make up. Figure 10 shows maps of where 3 elements (calcium, silicon and phosphate) are found within a region. Calcium, silicon and phosphate were chosen as calcium and phosphate are found in bone mineral, as well as in the CLM bio-glass scaffold in combination with silicon. Thus, they give us an insight into the scaffold behavior. The outer surfaces are now calcium and phosphorus rich, with the center of the struts demonstrating minimal CaP. It is possible that these elements are dissolving out and participating in the bone forming/deposition processes.

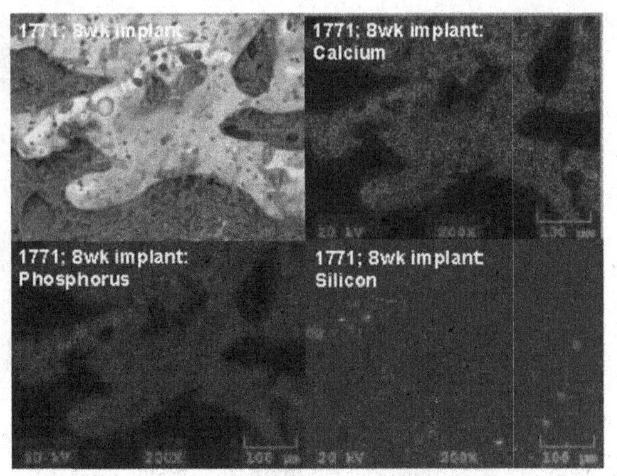

Figure 10. EDX images of mineral distributions.

CONCLUSION

There is a demand for the development of improved biomaterials to be used as scaffolds in bone and cartilage tissue engineering. Bioactive glass fibrous scaffolds with cross-linked microstructure may meet this demand, since bioactive glass fibers possess osteoconductivity, provide high strength and offer the possibility to develop new materials and manufacturing methods. A highly porous three-dimensional fibrous cross-linked microstructure network can be obtained. The porosity of a bioactive glass network not only noticeably increases the total reacting surface of the glass, but also serves as a framework for tissue ingrowth. This type of material may therefore prove useful as scaffolds for tissue engineering applications, for filling bone defects after trauma, infection and surgery.

ACKNOWLEDGMENTS

We wish to thank contributing team members at BIO2 Technologies, including Art O'Dea and Leonard Newton. This work was supported in part by BIO2 Technologies and the Department of Veterans Affairs.

REFERENCES

[1]D. W. Hutmacher, "Scaffolds in Tissue Engineering Bone and Cartilage," Biomaterials, 21, 2529–43 (2000).
[2]L. L. Hench, "Bioceramics," J. Am. Ceram. Soc., 81, 1705–28 (1998).
[3]M. M. Stevens, "Biomaterials for Bone Tissue Engineering," Mater. Today, 11, 18–25 (2008)
[4]J. R. Jones, E. Gentleman and J. Polak, "Bioactive Glass Scaffolds for Bone Regeneration," Elements 3[6], 393-399 (2007
[5]L. J. Bonassar and C.A. Vacanti, "Tissue Engineering: The First Decade and Beyond". J Cell Biochem. Suppl. 30–31:297–303 (1998).
[6]M. A. Meyers, P. Chen, A. Y. Lin, and Y. Seki, "Biological Materials: Structure and Mechanical Properties," Prog. Mater. Sci., 53, 1-206 (2008).
[7]B. Zuberi, J. J. Liu, S. C. Pillai, J. G. Weinstein, A. G. Konstandopoulos, S. Lorentzou, and C. Pagoura, "Advanced High Porosity Ceramic Honeycomb Wall Flow Filters" SAE Paper 2008-01-0623 (2008).

[8]J. J. Liu, R. A. Dahl, T. Gordon, and B. Zuberi, "Use Of Ceramic Microfibers To Generate A High Porosity Cross-linked Microstructure In Extruded Honeycombs," 33rd International Conference & Exposition on Advanced Ceramics & Composites, (2009)
[9]M Brink, "Bioactive glasses with a Large Working Range" Doctoral thesis, Åbo Akademi University, Turku, Finland; 1997.
[10]M. Brink, T. Turunen, R. Happonen, and A. Yli-Urppo, "Compositional Dependence of Bioactivity of Glasses in the System Na2O-K2O-MgO-CaO-B2O3-P2O5-SiO2," J. Biomed. Mater. Res., 37, 114–21 (1997)
[11]M. Brink, "The Influence of Alkali and Alkaline Earths on the Working Range for Bioactive Glasses," J Biomed. Mater. Res., 36:109–17 (1997).
[12]E. Pirhonen, M. Loredana, and J. Haapanen, "Porous Bioactive 3-D Glass Fiber Scaffolds for Tissue Engineering Applications Manufactured by Sintering Technique," Bioceramics 15. (2002)
[13]E. Pirhonen "Porous Bioactive 3-D Glass Fiber Scaffolds For Tissue Engineering Applications," Key Engineering Materials Vols. 240-242, 237-240 (2003).
[14]R. F. Brown, D. E. Day, T. E. Day, S. Jung, M. N. Rahaman, and Q. Fu, "Growth and Differentiation of Osteoblastic Cells on 13–93 Bioactive Glass Fibers and Scaffolds," Acta Biomater., 4, 387–96 (2008).
[15]Q. Fu, M. N. Rahaman, B. S. Bal, W. Huang, and D. E. Day, "Preparation and Bioactive Characteristics of a Porous 13–93 Glass, and Fabrication into the Articulating Surface of a Proximal Tibia," J. Biomed. Mater. Res., 82A, 222–9 (2007).
[16]Q. Fu, M. N. Rahaman, B. S. Bal, R. F. Brown, and D. E. Day, "Mechanical and In Vitro Performance of 13-93 Bioactive Glass Scaffolds Prepared by a Polymer Foam Replication Technique," Acta Biomater., 4, 1854-64 (2008).
[17]T. Kokubo, H. Kushitani, S. Sakka, T. Kitsugi, T. Yamamuro, "Solutions Able To Reproduce In Vivo Surface-Structure Changes in Bioactive Glass–Ceramic A–W," J. Biomed. Mater. Res. 24, 721–734 (1990).
[18]T. Kokubo and H. Takadama, "How Useful Is SBF in Predicting In Vivo Bone Bioactivity?," Biomaterials, 27 ,2907–2915 (2006)
[19]I. D. Xynos, M. J. Hukkanen, J. J. Batten, L. D. Buttery, L. L. Hench, J. M. Polak, "Bioglass 45S5 Stimulates Osteoblast Turnover and Enhances Bone Formation in Vitro. Implications and Applications For Bone Tissue Engineering" Calcif. Tissue Int. 67, 321–9 (2000).
[20]L. L. Hench, R. J. Splinter, W. C. Allen, and T. K. Greenlee Jr., "Bonding Mechanisms at the Interface of Ceramic Prosthetic Materials," J. Biomed. Mater. Res., 2, 117–41 (1971).
[21]M.A. De Diego, N. J. Coleman, and L.L. Hench, "Tensile Properties of Bioactive Fibres for Tissue Engineering Applications," J. Biomed. Mater. Res. (Appl. Biomater.), 53:199–203 (2000).
[22]D. L. Wheeler, K. E. Stokes, R. G. Hoellrich, D. L. Chamberland, and S. W. McLoughlin, "Effect of Bioactive Glass Particle Size on Osseous Regeneration of Cancellous Defects," J. Biomed. Mater. Res., 41, 527–33 (1998).

SELECTIVE LASER SINTERED Ca-P/PHBV NANOCOMPOSITE SCAFFOLDS WITH SUSTAINED RELEASE OF rhBMP-2 FOR BONE TISSUE ENGINEERING

Bin Duan [1], William W. Lu [2], Min Wang [1,*]
[1] Dept. of Mechanical Engineering, The University of Hong Kong, Pokfulam Road, Hong Kong
[2] Dept. of Orthopaedics & Traumatology, The University of Hong Kong, Sassoon Road, Hong Kong
[*] Email: memwang@hku.hk

ABSTRACT

Ca-P/PHBV nanocomposite scaffolds for bone tissue engineering were fabricated via selective laser sintering. The surface modification of Ca-P/PHBV scaffolds was conducted firstly by physical entrapment of gelatin. Heparin was then immobilized on gelatin-modified scaffolds through covalent conjugation. Human umbilical cord derived mesenchymal stem cells (hUC-MSCs) were seeded onto the scaffolds. Compared to non-modified scaffolds, heparin-immobilized scaffolds exhibited higher cell proliferation at the early stage of cell culture. hUC-MSCs became confluent after 21 day culture on scaffolds and covered the whole scaffold surface, strongly adhering to the scaffolds. Recombinant human bone morphogenetic protein (rhBMP)-2 was loaded onto scaffolds with or without surface modification and its *in vitro* release behavior was studied. An initial burst release of rhBMP-2 was observed for both types of scaffolds. However, the immobilization of heparin on the surface of Ca-P/PHBV scaffolds not only provided a means to protect the rhBMP-2 but also improved its sustained release. Surface modified scaffolds loaded with rhBMP-2 promoted significantly higher ALP activity of hUC-MSCs than the scaffolds with simple adsorption of rhBMP-2. The strategy of combining advanced scaffold fabrication technology, nanocomposite and growth factor delivery is promising for bone tissue regeneration.

INTRODUCTION

From the material point of view, human bones are a nanocomposite material consisting of an organic matrix (mainly collagen) and inorganic nanofillers (mainly bone apatite) which are inserted in a parallel way into the collagen fibrils.[1] Therefore, nanocomposites consisting of a biodegradable polymer and a nano-sized bioactive bioceramic such as hydroxyapatite (HA) or Bioglass®, which mimic the hierarchical structure of bone, are promising biomaterials for bone tissue repair or regeneration.[2, 3] In bone tissue engineering, controlling the macro- and micro-architecture of tissue engineering scaffolds and hence achieving customized designs of the scaffolds with complex anatomic shapes are also important for the clinical success of the scaffolds. In order to have an extensive and detailed control over scaffold architecture, designed manufacturing techniques, usually referring to as solid free-form fabrication (SFF) or rapid prototyping (RP) technologies in the general manufacturing sector of the industry, are investigated and developed for scaffold fabrication in the tissue engineering field.[4, 5] As a family member of RP technologies, the selective laser sintering (SLS) technique employs a CO_2 laser to selectively sinter thin layers of powdered polymers or their composites, forming solid three-dimensional porous scaffolds for tissue engineering applications.[6, 7] Some researchers have tried to incorporate nano-sized bioactive ceramics in RP-produced scaffolds and the incorporation was normally performed by dry-blending of polymer granules with bioceramic powders.[8] Nanocomposite scaffolds with controlled architecture, better osteoconductivity and improved properties such as surface properties and degradation rate could therefore be obtained.

It has been often reported that scaffold alone may not be sufficient to cause spontaneously healing and regeneration of functional body tissues due to a lack of factors that would promote cell proliferation and differentiation.[9, 10] These factors include hormone, proteins such as cytokines and growth factors and they are responsible for providing cell signals that can prompt specific cell behaviors and functions. Although growth factors play important roles in harnessing and controlling cellular functions in tissue regeneration, their short half-lives and rapid degradation, relatively large sizes, slow tissue penetration and potential toxicity at systemic levels leading to a long time for tissue to respond have limited their direct therapeutic applications.[11] Therefore, an effective and sustained delivery of growth factors such as basic fibroblast growth factor (bFGF), vascular endothelial growth factors (VEGF) and bone morphogenetic proteins-2 (BMP-2) at the target site are of significant importance. One delivery method is to encapsulate growth factors in the scaffolds.[12, 13] Another strategy is to immobilize growth factors onto the scaffold surface. The carriers and delivery methods selected for the sustained delivery of growth factors play a vital role in the bone regeneration process. Heparin, a sulfated polysaccharide, is known to have the binding affinity with a number of growth factors and thus capable of blocking the degradation of proteins and of prolonging the release time.[14, 15] In the presence of heparin, degradation of BMP-2 was blocked and the half-life of BMP-2 in the culture medium was prolonged by nearly 20-fold.[16] Heparin-conjugated scaffolds loaded with BMP-2 was found to increase the ectopic bone formation, indicating that the sustained and controlled delivery of BMP-2 increased the bone regenerative efficacy of BMP-2.[17] In addition, the incorporation of heparin into biomaterials was reported to have increased the biocompatibility and cell adhesion and growth.[18, 19]

For cell sources in bone tissue regeneration, the most widely investigated and used cells are osteoblasts and bone marrow derived mesenchymal stem cells (BM-MSCs). Primary osteoblasts or osteoprogenitor cells isolated from patients themselves do not raise ethical concerns and have no problems of immune rejection. However, it is time-consuming to obtain sufficient cells after cell isolation and expansion because the expansion rate is low. In certain bone-related diseases, patients' own osteoblasts may not be appropriate for transplantation because their protein expression is under the expected values.[20] BM-MSCs have the ability to differentiate into multilineages including osteoblasts, adipocytes, chondrocytes, etc.[21] However, the cell harvesting procedure is invasive and painful for donors and the multipotency of BM-MSCs is highly age dependent. Recently, much attention has been given to human umbilical cord derived mesenchymal stem cells (hUC-MSCs) due to their availability, noninvasive harvesting procedure and no serious ethical concerns.[22, 23]

In this paper, our research on integrating an advanced manufacturing technique suitable for complex structure scaffolds, biomimetic nanocomposite approach and controlled growth factor delivery for bone tissue engineering is reported. Calcium phosphate (Ca-P)/poly(hydroxybutyrate-co-hydroxyvalerate) (PHBV) nanocomposite scaffolds based on Ca-P/PHBV nanocomposite microspheres were fabricated via SLS. The nanocomposite scaffolds were subsequently subjected to surface modification with the entrapment of gelatin and immobilization of heparin. Recombinant human bone morphogenetic protein-2 (rhBMP-2) was incorporated in the nanocomposite scaffolds using its binding affinity with heparin which also controlled the rhBMP-2 release behavior. Local and sustained release of rhBMP-2 directed the osteogenesis differentiation of hUC-MSCs.

MATERIALS AND METHODS

Synthesis of Ca-P nanoparticles and fabrication of Ca-P/PHBV nanocomposite microspheres

Osteoconductive Ca-P nanoparticles were produced in-house by rapid mixing of $Ca(NO_3)_2 4H_2O$ (Uni-Chem, Orientalab, China) acetone solution with $(NH_4)_2HPO_4$ (Uni-Chem, Orientalab, China) aqueous solution.[24] With sizes in the range of 10-30 nm, the Ca-P nanoparticles obtained were amorphous and had a Ca:P molar ratio of about 1.5 which is similar to that of tricalcium phosphate, a commonly used bioactive and biodegradable bioceramic for bone tissue repair. A PHBV copolymer (12 mol% 3-hydroxyvalerate, ICI, UK) was used for Ca-P/PHBV nanocomposite microspheres which were fabricated using a solid-in-oil-in-water (S/O/W) emulsion solvent evaporation method.[25] The Ca-P/PHBV nanocomposite microspheres had an average diameter of 46.34

μm, as measured by a particle sizer (Mastersizer 2000 instrument, Malvern, UK), and contained 12.9 wt% of Ca-P, as determined through thermogravimetric analysis (TGA).

Design and fabrication of three-dimensional scaffolds

A rod-shaped scaffold model with a three-dimensional periodic porous architecture, as shown in Figure 1A, was designed using SolidWorks®. To improve the quality of sintered scaffolds and facilitate scaffold handling, a solid base (L W H=8.4 8.4 3 mm^3) was incorporated in the scaffold design. The scaffold model consisted of a repeating array of struts with a diameter of 1.0 mm in the three principal directions and the distance between each strut was set at 1.8 mm. The diameter of designed scaffold was 7.4 mm. The design was then exported into an STL format and transferred to a modified SLS machine (Sinterstation® 2000 system; 3D Systems, Valencia, CA, USA) for constructing scaffolds via SLS. According to our previous SLS optimization investigations, the laser power was set to be 15 W for Ca-P/PHBV nanocomposite scaffolds. The scan spacing and layer thickness was chosen to be 0.1 mm and 0.1 mm, respectively. The part bed temperature, scan speed and roller speed were fixed at 35 C, 1257 mm/s, and 127 mm/s, respectively.

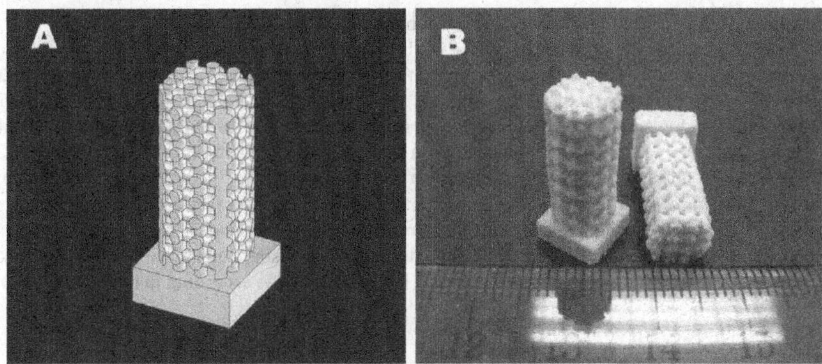

Figure 1. (A) Rod-shaped scaffold model; and (B) Ca-P/PHBV scaffolds fabricated via SLS

Scaffold surface modification

The surface modification of Ca-P/PHBV nanocomposite scaffolds was conducted in two steps: (1) physical entrapment of gelatin, and (2) heparin immobilization. Gelatin (Sigma, USA) was first physically entrapped by immersing sintered Ca-P/PHBV scaffold samples (two-layered scaffolds, ~70 mg/each scaffold) in a gelatin solution, which was made by a miscible mixture of 2,2,2-trifluoroethanol (TFE; Acros, Belgium) and distilled water (TFE : water = 30 : 70), at room temperature for 6 h. After the treatment, the surface modified scaffolds were rinsed in distilled water three times to remove non-entrapped gelatin and then dried at room temperature. The amount of entrapped gelatin on the scaffold surface was quantitatively determined by bicinchoninic acid (BCA) kit assay (Pierce, Rockford, IL, USA). For heparin immobilization, gelatin-entrapped Ca-P/PHBV scaffolds were firstly prewetted in a 2-(N-morpholino)ethanesulfonic acid (MES) buffer solution (0.1 M, pH 5.6; Fluka, USA) for 30 min at room temperature. 3 mg of heparin (Mw = 17,000, activity 170USP units/mg; Sigma, USA) were dissolved in 1 ml MES buffer solution (0.1 M, pH5.6) containing 2 mg N-hydroxysuccinimide (NHS; Sigma, USA) and 1.2 mg 1-ethyl-3-(3-dimethylaminopropyl)-carbodiimide hydrochloride (EDC; Sigma, USA). After 4 h activation at room temperature, the prewetted scaffolds were soaked in the activated heparin solution in MES buffer for another 4 h and then extensively washed with phosphate buffered saline (PBS) and dried overnight at room temperature.

Characterization of scaffolds

The surface morphology of sintered Ca-P/PHBV nanocomposite scaffolds before and after gelatin entrapment and heparin immobilization, respectively, were examined using a scanning electron microscope (SEM, Hitachi S-3400N, Japan). Due to the high porosity of porous scaffolds, instead of using sintered scaffolds, the hydrophilicity of surfaces before and after surface modification was studied through contact angle measurement of PHBV films which were produced using the solvent-casting method. Briefly, PHBV-chloroform solution was cast into a glass Petri dish and placed in a fume hood at room temperature for the slow evaporation of chloroform. After further vacuum drying, PHBV films were collected and cut into square samples of the dimensions 1 cm 1 cm. The surface modification of solvent-cast PHBV films adopted the same procedure as described in the previous section. Water contact angles were measured at room temperature with an image analyzing system using the sessile drop technique.

Incorporation of rhBMP-2 and its *in vitro* release behavior

For the incorporation of rhBMP-2, 5 μg of rhBMP-2 (Shanghai Rebone Biomaterials Co. Ltd., China) were loaded onto each scaffold sample (two-layered scaffolds) with and without surface modification by dripping 50 μl rhBMP-2 solutions with 20 mM glacial acetic acid and 0.1% bovine serum albumin (BSA). After incubation for 1 h at room temperature, the rhBMP-2 loaded scaffolds were placed in centrifugal tubes and suspended in 2 ml of PBS containing 0.1% BSA and 0.02% sodium azide. The tubes were sealed and placed in a shaking water bath (SW22, Julabo, Germany), which was maintained at 37 C and shaken horizontally at 30 rpm. At preset times up to 28 days, all supernatants were collected and another 2 ml fresh medium was replenished for each tube. The amount of released rhBMP-2 in the collected medium was determined by using human BMP-2 enzyme-linked immuno-adsorbent assay (ELISA) development kit (PeproTech, USA) according to the manufacturer's protocol.

Culture of hUC-MSCs

The frozen hUC-MSCs were firstly thawed rapidly and then diluted slowly into warm basal medium containing DMEM/F12 (Invitrogen, USA) with 20% fetal bovine serum, 100 U/ml penicillin–streptomycin, 4 μg/ml fungizone, and 2 mM L-glutamine by adding dropwise the cells. After thawing, the culture medium was firstly refreshed after 1 day culture. Then, the medium was replaced every 2 days and cultures were maintained in a humidified incubator at 37 °C with 5% CO_2. After reaching about 80% confluence, the cells were digested and subcultured using 0.25% (w/v) trypsin-ethylenediamineteraacetic acid (EDTA) (Invitrogen, USA). The resulting cells in a suspension were then seeded separately onto two-layered scaffold samples which had been sterilized by ^{60}Co γ-irradiation at a dose of 25 kGy. For control experiments, polystyrene tissue culture plates (TCPs) were used. Cell seeding onto samples was conducted by dripping 100 μl cell suspension with the cell density of 5×10^6 cells/ml in 24-well plates and then the culture wells were filled with 1 ml of culture medium after 2 h in the incubator.

Cell morphology

After 21 day culture in osteogenic medium (the basal medium supplemented with 10 nM dexamethasone, 0.05 mM L-ascorbic acid, and 0.01 mM β-glycerophosphate.), cell-scaffold constructs were harvested, washed twice with PBS, and subsequently fixed with 2.5% glutaraldehyde at 4 °C for 4 h. After washing with cacodylate buffer containing 0.1 M sucrose, they were dehydrated through a series of graded alcohol solutions and dried in a critical point dryer using liquid carbon dioxide as the transition fluid. The samples were then glued onto SEM stubs and sputter-coated with a thin layer of gold for observation under SEM.

Osteogenic differentiation

For determining the osteogenic capacity of hUC-MSCs, the cells were cultured in the osteogenic medium. In order to study the effect of rhBMP-2 on the osteogenic differentiation of hUC-MSCs, 1000 ng/ml rhBMP-2 was also added into the osteogenic medium. The medium was refreshed every two to three days during the 21 day study period. After 14 days of osteogenic induction, hUC-MSCs were stained with the alkaline phosphatase leukocyte kit (Sigma, USA) according to the manufacturer's protocol. The calcium deposition was stained with 1% alizarin red S after fixation by 10% neutralized formalin after 21 day culture.

For studying the osteogenic differentiation of hUC-MSCs on nanocomposite scaffolds, four groups of scaffold samples were involved, namely, bare Ca-P/PHBV scaffolds, Ca-P/PHBV scaffolds with simple adsorption of rhBMP-2, Ca-P/PHBV scaffolds with surface modification, and surface modified Ca-P/PHBV scaffolds with rhBMP-2. For the two types of Ca-P/PHBV scaffolds with rhBMP-2, 5 μg rhBMP-2 were loaded onto each scaffold (with or without surface modification) by dripping 50 μl rhBMP-2 solutions with 20 mM glacial acetic acid and 0.1% BSA.

Alkaline phosphatase (ALP) activity

The ALP activity was measured after 7, 14 and 21 day cell culture after osteogenic induction. Cell-scaffold constructs were rinsed twice with PBS to remove non-adhering cells. They were then treated with 200 μl trypsin solution for several minutes to detach cells. Subsequently, 500 μl culture medium was added to end the digestion. After centrifugation at 5000 rpm for 5 min and further wash with PBS, the obtained cells were kept in 500 μl lysis buffer containing 0.1% (v/v) Triton X-100, 1 mM $MgCl_2$, and 20 mM Tris, which was followed by a freezing and thawing process to further disrupt the cell membranes. A 50 μl lysate was mixed with 200 μl ALP substrate solution containing *p*-nitrophenyl phosphate (pNPP) (Sigma, USA) at 37 °C for 30 min. The reaction was stopped by the addition of 50 μl of 3 N NaOH and then the production of *p*-nitrophenol in the presence of ALPase was measured by monitoring the absorbance of the solution at a wavelength of 405 nm using a microplate reader. The total protein content was determined using BCA assay kit with BSA as a standard and the ALP activity was expressed as μmol of *p*-nitrophenol formation per minute per milligram of total proteins (μmol/min/mg protein).

Statistical analysis

All quantitative data were obtained from triplicate samples and expressed as the mean ± standard deviation (SD). Statistical analysis was performed using ANOVA with a Scheffé test. A value of $p<0.05$ was considered to be statistically significant and $p<0.01$ remarkably significant.

RESULTS

Fabrication of three-dimensional Ca-P/PHBV nanocomposite scaffolds

The SLS technique employs a CO_2 laser to selectively sinter thin layers of powdered polymers or polymeric composite materials, forming solid 3D objects. Ca-P/PHBV nanocomposite microspheres were thus fabricated and served as raw materials for the scaffold sintering process. The Ca-P nanoparticles with sizes in the range of 10-30 nm were firstly synthesized in-house. The obtained Ca-P nanoparticles were amorphous and had a Ca:P molar ratio of about 1.5. Ca-P/PHBV nanocomposite microspheres based on Ca-P nanoparticles and biodegradable PHBV matrix were fabricated using the

S/O/W emulsion solvent evaporation method. The Ca-P/PHBV microspheres had an average diameter of 46.34 μm, which were of the appropriate sizes for selective laser sintering. This new manufacturing strategy and route ensure good distribution of bioceramic nanoparticles in the composite scaffolds. Another advantage of the current approach is that some drugs or even biomolecules could also be incorporated into the microspheres (and hence the scaffolds after SLS) in order to form multifunctional nanocomposite scaffolds and improve their bioactivity.[26] Using appropriate SLS parameters, three-dimensional rod-shaped Ca-P/PHBV scaffolds (Figure 1B) were fabricated for the surface modification study and for the rhBMP-2 loading and release study. The sintered scaffolds had intact structure and possessed good handling stability. The porosity of the scaffold model was calculated to be 53.5% using the SolidWorks® software, while the actual porosity values of sintered scaffolds were 61.8 1.2%, which was higher than 53.5% due to the presence of micropores on the surface of scaffold struts (Figure 2B). These scaffolds also had a fully interconnected porous structure which is vital for the diffusion of nutrients and gases and for the removal of metabolic waste resulting from the activity of the cells grown in the scaffolds and can facilitate *in vivo* vascularization.

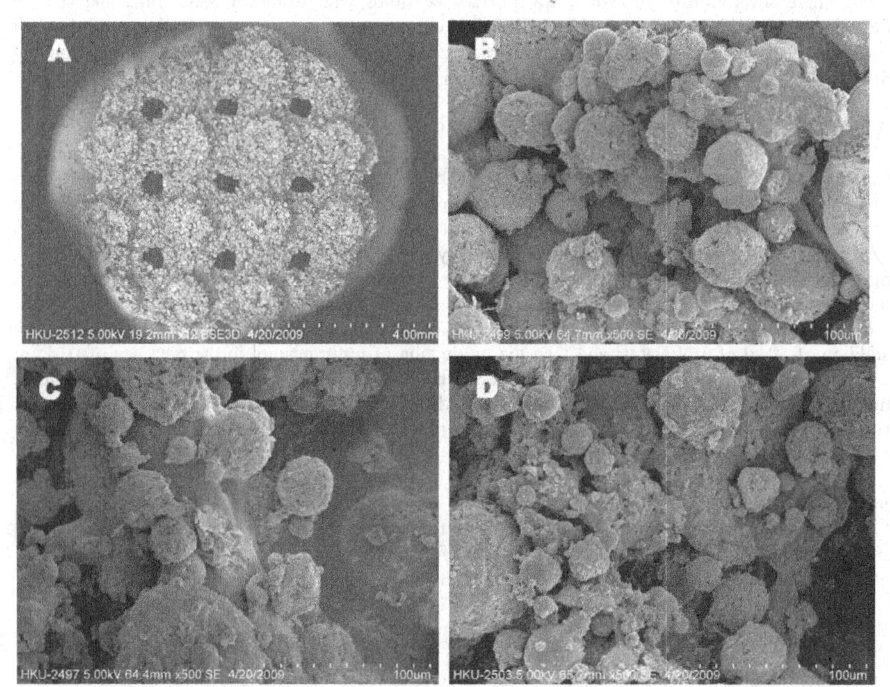

Figure 2. SEM micrographs of sintered Ca-P/PHBV nanocomposite scaffold: (A, B) before surface modification; (C) after gelatin entrapment; and (D) after further heparin immobilization. (A: general view; B, C, D: close view.)

Surface modification of Ca-PHBV nanocomposite scaffolds

For surface modification, two steps were involved: (1) physical entrapment of gelatin, and (2) heparin immobilization. The entrapment of gelatin required a gelatin solution based on a miscible mixture of solvent and non-solvent for the PHBV matrix. During the entrapment process, the PHBV matrix swelled but did not dissolve in the gelatin solution. The gelatin molecules diffused into the swollen surface of PHBV and were entangled with PHBV molecules on the surface. After the scaffolds were removed from the gelatin solution and immersed into distilled water, which is a non-solvent for the polymer, the surface rapidly shrank and gelatin molecules on the polymer surface were entrapped and immobilized.[27] The amount of entrapped gelatin was measured by BCA method, which is an effective way to determine the amount of immobilized proteins on the surface of biomaterials. The amount of entrapped gelatin was measured to be 1955.8 62.1 μg per scaffold. The entrapment of gelatin molecules introduced on strut surfaces amino groups, which could be used for the conjugation of heparin. Heparin was subsequently immobilized onto the gelatin-entrapped scaffold surface based on standard carbodiimide chemistry.[28, 29] The amount of immobilized heparin was determined to be 41.78 0.39 μg per scaffold using the toluidine blue method which is a facile way to verify and determine the presence of conjugated heparin based on the formation of a strong purple complex with the sulfated polysaccharide.[30]

SEM micrographs of scaffolds before and after surface modification are shown in Figure 2. No significant difference in scaffold morphology was observed after gelatin entrapment and heparin immobilization, respectively. Solvent-cast PHBV films instead of scaffolds were used for water contact angle measurements due to the high porosity of scaffolds. The water contact angle for PHBV films before surface modification was 93.61 1.85 , while it decreased to 75.12 3.79 after gelatin entrapment. The physical entrapment of gelatin improved the hydrophilicity of PHBV (and hence the scaffold). After further immobilization of heparin, the water contact angle decreased slightly to 72.06 0.89 . The main purpose of heparin immobilization is to provide specific affinity between heparin immobilized scaffold and growth factors (to be incorporated) and to control the release profile of bound growth factors in a sustained manner so as to improve angiogenesis and/or osteogenesis.

Incorporation of rhBMP-2 and its *in vitro* release behavior

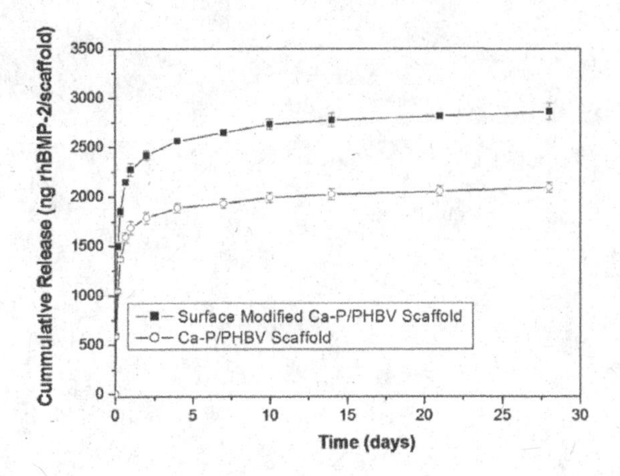

Figure 3. Cumulative release of rhBMP-2 from Ca-P/PHBV scaffolds with and without surface modification.

As described earlier, rhBMP-2 was loaded onto two types of Ca-P/PHBV nanocomposite scaffolds: bare scaffolds, and scaffolds with physical entrapment of gelatin and heparin immobilization (viz., surface modified scaffolds). The *in vitro* release profiles of rhBMP-2 from these two types of Ca-P/PHBV scaffolds were determined using rhBMP-2 ELISA kit and are shown in Figure 3. Both types of scaffolds displayed an initial burst release with a subsequent sustained release over the 28 day test period. Approximately 2276.11 66.95 ng rhBMP-2 of cumulative release from surface modified Ca-P/PHBV scaffolds were observed after the 1 day release time, while about 1686.71 66.07 ng rhBMP-2 of cumulative release were found for scaffolds without surface modification with the same release time. The surface modified scaffolds exhibited higher initial burst release but delivered rhBMP-2 in a more sustained manner than scaffolds without surface modification. Similar sustained release behaviors for growth factors such as BMP-2, VEGF, bFGF and platelet-derived growth factor (PDGF) were observed for heparin-conjugated scaffold or microsphere delivery systems.[31, 32] After the 28 day test period, totally about 2860.79 82.54 ng rhBMP-2 were released from surface modified Ca-P/PHBV scaffolds, which were significantly more than the amount released from scaffolds without surface modification (2095.90 51.48 ng).

Osteogenic differentiation of hUC-MSCs

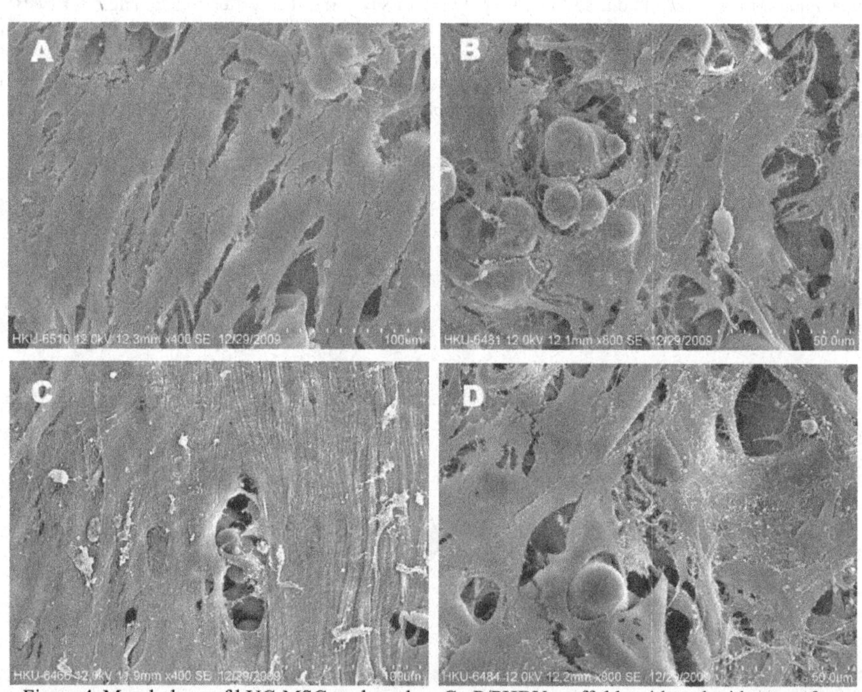

Figure 4. Morphology of hUC-MSCs cultured on Ca-P/PHBV scaffolds with and without surface modification for 21 days: (A, B) scaffold without surface modification; (C, D) scaffold with surface modification. (A, C: general view; B, D: close view)

Figure 4 shows representative hUC-MSC cell morphology on Ca-P/PHBV nanocomposite scaffolds with and without surface modification. Obviously, hUC-MSCs became confluent after 21 day culture on the sintered scaffold substrates and covered wholly the scaffold surface. From the close views of cell-scaffold constructs (Figure 4B and D), it could be seen that hUC-MSCs spread well and strongly adhered and anchored themselves onto the microspheres of sintered scaffold struts.

After cell culture in the osteogenic medium, the ALP activity of cells on cell-seeded scaffolds, namely, Ca-P/PHBV, surface modified Ca-P/PHBV scaffolds, Ca-P/PHBV scaffolds with rhBMP-2, and surface modified Ca-P/PHBV scaffolds with rhBMP-2, were analyzed and the results are shown in Figure 5. The ALP activity was different for the four scaffold groups. For each type of scaffolds, the ALP activity generally increased between day 7 and 14 and reached maximum at around day 14, indicating osteogenic differentiation. Surface modified scaffolds loaded with rhBMP-2 exhibited the highest enzyme activity and showed significantly higher ALP expression than scaffolds with simple adsorption of rhBMP-2 on day 7 and day 21 ($p<0.05$ on day 7 and $p<0.01$ on day 21). Without the presence of rhBMP-2, no significant difference in ALP activity was observed between scaffolds with and without surface modification. The ALP activity of cells on all scaffold samples, as well as on tissue culture plates, decreased on day 21 because ALP is an early marker for osteogenic differentiation and its activity usually peaks at the early stage of *in vitro* experiments.

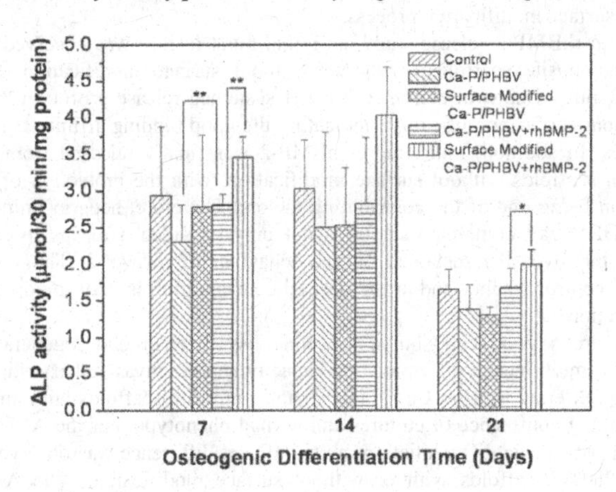

Figure 5. ALP activity of hUC-MSCs seeded on different scaffolds and also tissue culture plate (the control) for different culture times. (*$p < 0.05$; **$p < 0.01$)

DISCUSSION

In the current study, we pioneered producing and using nanocomposite microspheres as raw materials in the SLS process for fabricating three-dimensional osteoconductive bone tissue engineering scaffolds. This strategy is unlike previously reported research by others in which bioceramic particles were dry-blended with polymer granules and the blends were then used directly for SLS to form composite scaffolds. For SLS, raw materials must be in the powder form and should be of appropriate particle sizes and have good flowability for spreading on the part bed of the SLS machine. Therefore, materials with particle sizes in the range of 10-150 μm are preferred. In this study, all microspheres fabricated were of suitable sizes for SLS and Ca-P/PHBV nanocomposite scaffolds were successfully fabricated via SLS according to the designed scaffold model. Using nanocomposite microspheres has several advantages over using dry-blended raw materials in constructing bone tissue engineering scaffolds through SLS. Although the "microsphere" route takes longer time in preparing the raw material for SLS than the "dry-blending" route, considerations for the homogeneity of bioceramic particles in scaffolds, scaffold quality, scaffold mechanical properties, etc. outweigh the manufacture time factor.

Even though the incorporation of osteoconductive Ca-P nanoparticles in microspheres and hence scaffolds improves the cytocompatibility and bioactivity of polymeric scaffolds, insufficient osteoinductivity and osteointegration could still result in the failure of a scaffold-based bone tissue engineering strategy. In order to further enhance the functionality of tissue engineering scaffolds, biological molecules, especially growth factors, can be incorporated in scaffolds. In the current study, gelatin was first physically entrapped onto the surface of Ca-P/PHBV nanocomposite scaffolds. Based on carbodiimide chemistry, the entrapped gelatin then provided amino groups for the conjugation of heparin. Heparin is known to provide the binding site for growth factors and thus can control the release of a growth factor in a sustained manner. The surface modification by entrapment of gelatin and immobilization of heparin significantly improved the wettability of solvent-cast PHBV films, suggesting the improvement of wettability of surface modified Ca-P/PHBV nanocomposite scaffolds. However, the morphology and mechanical properties of Ca-P/PHBV nanocomposite scaffolds were not affected by the surface modification process.

In the *in vitro* rhBMP-2 release study, an initial burst release was observed for both types of Ca-P/PHBV nanocomposite scaffolds: with and without surface modification. Surface modified scaffolds with immobilized heparin exhibited a better sustained release profile in the release process than scaffolds without surface modification, indicating the good binding affinity between heparin and rhBMP-2 molecules. In addition, much more rhBMP-2 was finally released from surface modified scaffolds than from scaffolds without surface modification, with the protection of rhBMP-2 by the immobilized heparin being one of the contributing factors. Therefore, heparin immobilization on the surface of Ca-P/PHBV nanocomposite scaffolds not only provided a means to protect the loaded rhBMP-2 but also improved the sustained release behaviour of rhBMP-2. The sustained release of growth factors in a control manner and at the desired concentration is very important for stimulating bone tissue regeneration.

hUC-MSCs are a good cell source for bone tissue repair and regeneration due to their inexhaustible supply, multilineage differentiation capacity and noninvasive harvesting procedure. The hUC-MSCs could proliferate well on Ca-P/PHBV nanocomposite scaffolds with and without surface modification, exhibiting confluence of cultures and normal phenotype. For the ALP activity, which is an early marker for osteogenic differentiation, no significant difference was observed for hUC-MSCs cultured on Ca-P/PHBV scaffolds with or without surface modification. The ALP activity assay showed that the ALP level was significantly up-regulated on surface modified Ca-P/PHBV scaffolds loaded with rhBMP-2 as compared to scaffolds with simple adsorption of rhBMP-2 ($p<0.01$ at day 7 and $p<0.05$ at day 21). These results also indicated that the immobilization of heparin could facilitate the sustained release of rhBMP-2 and stimulate the osteogenic differentiation of hUC-MSCs.

CONCLUSIONS

The current study has demonstrated that it is feasible to integrate the SLS technique, nanocomposite material and sustained release of growth factor to form advanced bone engineering scaffolds. Good-quality, three-dimensional nanocomposite scaffolds based on Ca-P/PHBV nanocomposite microspheres could be fabricated via SLS. The surface of Ca-P/PHBV nanocomposite scaffolds could be modified for incorporating growth factors by gelatin entrapment and heparin immobilization. This surface modification improved the wettability of scaffolds and provided binding affinity between conjugated heparin and the growth factor rhBMP-2. The immobilization of heparin on the surface of Ca-P/PHBV nanocomposite scaffolds not only provided a means to protect the loaded rhBMP-2 but also improved its sustained release behaviour. The ALP activity of hUC-MSCs on surface modified scaffolds loaded with rhBMP-2 was significantly higher than that of the cells on scaffolds with simple adsorption of rhBMP-2 during the 21 day cell culture period. The creation of customized scaffolds with controlled architecture, osteoconductive property and sustained release of the osteogenic growth factor BMP-2offers great potential for bone tissue engineering.

ACKNOWLEDGEMENTS

This work was supported by a research grant from The University of Hong Kong (HKU) and by a GRF grant (HKU 7176/08E) from the Hong Kong Research Grants Council. The authors thank Dr. W.-L. Cheung of HKU for his advice on SLS and Prof. C.-S. Liu of East China University of Science and Technology, China, for providing rhBMP-2.

REFERENCES

[1]P. Fratzl, H.S. Gupta, E.P. Paschalis and P. Roschger, Structure and mechanical quality of the collagen-mineral nano-composite in bone, *J. Mater. Chem.*, **14**, 2115-23 (2004).

[2]B.M. Chesnutt, Y.L. Yuan, K. Buddington, W.O. Haggard and J.D. Bumgardner, Composite chitosan/nano-hydroxyapatite scaffolds induce osteocalcin production by osteoblasts in vitro and support bone formation in vivo, *Tissue Eng. Part A,* **15**, 2571-79 (2009).

[3]X. Li, G. Koller, J. Huang, L. Di Silvio, T. Renton, M. Esat, W. Bonfield and M. Edirisinghe, A novel jet-based nano-hydroxyapatite patterning technique for osteoblast guidance, *J. R. Soc. Interface*, **7**, 189-97 (2010).

[4]D.W. Hutmacher, M. Sittinger, M.V. Risbud, Scaffold-based tissue engineering: rationale for computer-aided design and solid free-form fabrication systems, *Trends Biotechnol*, **22**, 354-62 (2004).

[5]M.W. Naing, C.K. Chua, K.F. Leong and Y. Wang, Fabrication of customised scaffolds using computer-aided design and rapid prototyping techniques, *Rapid Prototyping J.*, **11** 249-59 (2005).

[6]J.M. Williams, A.Adewunmi, R.M. Schek, C.L. Flanagan, P.H. Krebsbach, S.E. Feinberg, S.J. Hollister and S. Das, Bone tissue engineering using polycaprolactone scaffolds fabricated via selective laser sintering, *Biomaterials*, **26**, 4817-27 (2005).

[7]W.Y. Yeong, N. Sudarmadji, H.Y. Yu, C.K. Chua, K.F. Leong, S.S. Venkatraman, Y.C.F. Boey and L.P. Tan, Porous polycaprolactone scaffold for cardiac tissue engineering fabricated by selective laser sintering, *Acta Biomater.*, **6**, 2028-34 (2010).

[8]S.J. Heo, S.E. Kim, J. Wei, Y.T. Hyun, H.S. Yun, D.H. Kim, J.W. Shin and J.W. Shin, Fabrication and characterization of novel nano- and micro-HA/PCL composite scaffolds using a modified rapid prototyping process, *J. Biomed. Mater. Res. Part A*, **89A**, 108-16 (2009).

[9]C.H. Chang, T.F. Kuo, C.C. Lin, C.H. Chou, K.H. Chen, F.H. Lin and H.C. Liu, Tissue engineering-based cartilage repair with allogenous chondrocytes and gelatin-chondroitin-hyaluronan tri-copolymer scaffold: a porcine model assessed at 18, 24, and 36 weeks, *Biomaterials*, **27**, 1876-88 (2006).

[10]Y.F. Zhang, C.T. Wu, T. Friis and Y. Xiao, The osteogenic properties of CaP/silk composite scaffolds, *Biomaterials*, **31**, 2848-56 (2010).

[11]J.R. Porter, T.T. Ruckh and K.C. Popat, Bone tissue engineering: a review in bone biomimetics and drug delivery strategies, *Biotechnol. Prog.*, **25**, 1539-60 (2009).

[12]J.A. Kanczler, P.J. Ginty, J.J.A. Barry, N.M.P. Clarke, S.M. Howdle, K.M. Shakesheff and R.O.C. Oreffo, The effect of mesenchymal populations and vascular endothelial growth factor delivered from biodegradable polymer scaffolds on bone formation, *Biomaterials*, **29**, 1892-1900 (2008).

[13]X.Q. Li, Y. Su, S.P. Liu, L.J. Tan, X.M. Mo and S. Ramakrishna, Encapsulation of proteins in poly(L-lactide-co-caprolactone) fibers by emulsion electrospinning, *Colloid Surf. B-Biointerfaces*, **75**, 418-24 (2010).

[14]I. Capila and R.J. Linhardt, Heparin - Protein interactions, *Angew. Chem.-Int. Edit.*, **41**, 391-412 (2002).

[15]I. Freeman and S. Cohen, The influence of the sequential delivery of angiogenic factors from affinity-binding alginate scaffolds on vascularization, *Biomaterials*, **30**, 2122-31 (2009).
[16]B.H. Zhao, T. Katagiri, H. Toyoda, T. Takada, T. Yanai, T. Fukuda, U.I. Chung, T. Koike, K. Takaoka and R. Kamijo, Heparin potentiates the in vivo ectopic bone formation induced by bone morphogenetic protein-2, *J. Biol. Chem.*, **281**, 23246-53 (2006).
[17]H. Lin, Y.N. Zhao, W.J. Sun, B. Chen, J. Zhang, W.X. Zhao, Z.F. Xiao and J.W. Dai, The effect of crosslinking heparin to demineralized bone matrix on mechanical strength and specific binding to human bone morphogenetic protein-2, *Biomaterials,* **29**, 1189-97 (2008).
[18]S.Y. Meng, M. Rouabhia, G.X. Shi, Z. Zhang, Heparin dopant increases the electrical stability, cell adhesion, and growth of conducting polypyrrole/poly(L,L-lactide) composites, *J. Biomed. Mater. Res. Part A,* **87A**, 332-44 (2008).
[19]D.H. Go, Y.K. Joung, S.Y. Park, Y.D. Park and K.D. Park, Heparin-conjugated star-shaped PLA for improved biocompatibility, *J. Biomed. Mater. Res. Part A,* **86A,** 842-48 (2008).
[20]A.J. Salgado, O.P. Coutinho, R.L. Reis, Bone tissue engineering: state of the art and future trends. *Macromol. Biosci.*, **4**, 743-65 (2004).
[21]A.J. Leo and D.A. Grande, Mesenchymal stem cells in tissue engineering, *Cells Tissues Organs*, **183**, 112-22 (2006).
[22]E. Zucconi, N.M. Vieira, D.F. Bueno, M. Secco, T. Jazedje, C.E. Ambrosio, M.R. Passos-Bueno, M.A. Miglino and M. Zatz, Mesenchymal stem cells derived from canine umbilical cord vein - A novel source for cell therapy studies, *Stem Cells Dev.*, **19**, 395-402 (2010).
[23]P. Huang, L.M. Lin, X.Y. Wu, Q.L. Tang, X.Y. Feng, G.Y. Lin, X.B. Lin, H.W. Wang, T.H. Huang and L. Ma, Differentiation of human umbilical cord Wharton's Jelly-derived mesenchymal stem cells into germ-like cells in vitro, *J. Cell. Biochem.*, **109**, 747-54 (2010).
[24]B. Duan, M .Wang, W.Y. Zhou and W.L. Cheung, Synthesis of Ca-P nanoparticles and fabrication of Ca-P/PHBV microspheres for bone tissue engineering applications, *Appl. Surf. Sci.*, **255**, 592-33 (2008).
[25]B. Duan, M. Wang, W.Y. Zhou, W.L. Cheung, Z.Y. Li, W.W. Lu, Three-dimensional nanocomposite scaffolds fabricated via selective laser sintering for bone tissue engineering, *Acta Biomater.*, **6**, 4495-05 (2010).
[26]B. Duan and M. Wang. Encapsulation and release of biomolecules from Ca-P/PHBV nanocomposite microspheres and three-dimensional scaffolds fabricated by selective laser sintering, *Polym. Degrad. Stabil.*, **95**, 1655-64 (2010).
[27]Z.H. Liu, Y.P. Jiao, Z.Y. Zhang and C.R. Zhou, Surface modification of poly(L-lactic acid) by entrapment of chitosan and its derivatives to promote osteoblast-like compatibility, *J. Biomed. Mater. Res. Part A,* **83A,** 1110-16 (2007).
[28]O. Jeon, S.J. Song, S.W. Kang, A.J. Putnam and B.S. Kim, Enhancement of ectopic bone formation by bone morphogenetic protein-2 released from a heparin-conjugated poly(L-lactic-co-glycolic acid) scaffold, *Biomaterials*, **28**, 2763-71 (2007).
[29]H.J. Chung, H.K. Kim, J.J. Yoon and T.G. Park, Heparin immobilized porous PLGA microspheres for angiogenic growth factor delivery, *Pharm. Res.*, **23**, 1835-41 (2006).
[30]B. Sun, B. Chen, Y.N. Zhao, W.J. Sun, K.S. Chen, J. Zhang, Z.L. Wei, Z.F. Xiao and J.W. Dai, Crosslinking heparin to collagen scaffolds for the delivery of human platelet-derived growth factor, *J. Biomed. Mater. Res. Part B,* **91B,** 366-72 (2009).
[31]O. Jeon, S.W. Kang, H.W. Lim, J.H. Chung and B.S. Kim, Long-term and zero-order release of basic fibroblast growth factor from heparin-conjugated poly(L-lactide-co-glycolide) nanospheres and fibrin gel, *Biomaterials,* **27,** 1598-1607 (2006).
[32]Y.C. Ho, F.L. Mi, H.W. Sung and P.L. Kuo, Heparin-functionalized chitosan-alginate scaffolds for controlled release of growth factor, *Int. J. Pharm.*, **376,** 69-75 (2009).

MICROBEAM X-RAY GRAIN AVERAGED RESIDUAL STRESS IN DENTAL CERAMICS

Hrishikesh A. Bale,[1] Nobumichi Tamura[2], Jay C. Hanan,[1]

[1]Oklahoma State University, Stillwater, OK,

[2]Advanced Light Source, Berkeley, CA.

ABSTRACT

Ceramic dental restorations consist of translucent porcelain and an underlying structural ceramic core. The maximum bite loads in service (<200 MPa, molars) are far below failure stresses. In spite of low stresses, ceramic dental restorations can undergo failures in their first year. Zirconia is widely accepted for cores due to toughness, compatibility, and aesthetics. Its tetragonal-to-monoclinic phase transformation produces compressive residual stresses in the vicinity of concentrated tensile stresses. For complex dental crown geometries, eliminating residual stresses is not trivial. Moreover, techniques interpreting internal stress states are lacking. Micro-diffraction using highly focused monochromatic X-ray beams provides residual stress. In addition, spatially resolved phase transformation maps in zirconia have been demonstrated. Stresses in the tetragonal phase were based on measured stress-free lattice parameters. In the neighborhood of transforming grains, tetragonal grains showed compressive residual stresses. Tensile residual stresses of 0.5 GPa in the core with monoclinic transformation were observed. This advances understanding of residual stresses and phase transformations in failure and reliability of dental restorations and related bio-ceramics.

INTRODUCTION

All-ceramic dental restorations have evolved as a preferred class of dental prosthesis compared to the metal-porcelain bonded restorations. The excellent esthetic quality and non-toxic nature of ceramics have made them popular. The all-ceramic restorations are primarily made up of a layered configuration consisting of a structural core and an esthetic translucent veneer. Structural cores are typically made from alumina or zirconia, whereas, the veneer is generally porcelain [1-7]. Veneering of the structural cores involves firing the two different materials at a temperature where the veneer fuses to the structural core. In spite of careful selection of materials to reduce mismatch in the coefficients of thermal expansion, substantial interfacial residual stresses still persist. Residual stresses, especially tensile in nature, that reside at concentrated sites are severely detrimental to the life of the restoration [8-11]. An understanding of the magnitude and distribution over a given component can be established at a high spatial resolution through micro-diffraction using high energy synchrotron X-ray beam. The diffraction patterns resulting from a highly focused monochromatic X-ray beam provide information on the biaxial residual stress in the material. In the current work, we present the use of biaxial stress measurements using the *in-situ* $\sin^2\psi$ technique of X-ray diffraction [12-13]. This well established technique has been utilized widely in macro-beam mode, with beam cross sections typically as wide as 1 mm and above [13-15]. Here, we present the method and initial results obtained from the technique adapted to micro-beams with a nominal cross section of 2 μm × 10 μm. Small beam cross-sections facilitate the acquisition of stress information from a subset of few grains instead of a grain averaged result over a larger area. Especially considering the smaller grain sizes (< 1 μm) in dense ceramics such as alumina and zirconia, the technique produces biaxial stress results at a higher resolution. The results can also be averaged to reproduce what is seen with larger beams. High spatial resolution biaxial stress measurements from sectioned area of zirconia-core dental crown with a porcelain veneer are presented.

METHOD

Specimen Fabrication

Specimens comprised of dental crowns made in a clinical dental laboratory (Marotta Dental Studio, Huntington, NY). Preparation of the dental crowns is based on precision CNC milling of the structurally strengthening ceramic core (typically made of ZrO_2 or Al_2O_3) followed by firing a top porcelain veneer on the core. The detailed procedure is mentioned elsewhere [1, 16].

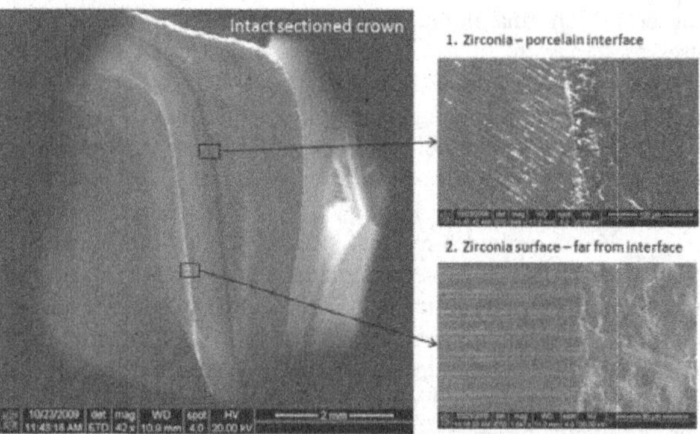

Figure 1 SEM micrographs of interface(1) and wet-cut core surface (2). SEM showed a good bonding between porcelain and zirconia. (2) shows the uneven striations on the surface of zirconia caused in the process of cutting with a diamond saw.

Residual stress measurements were carried out on crowns with zirconia cores. Initial experiments were performed to measure the interface residual stress by the X-ray beam penetrating through the porcelain of an un-cut intact crown. However due to the high absorption through the porcelain layer, along with irregular porcelain thickness, sufficiently bright diffraction peaks were not achievable. Due to this constraint, bi-axial surface measurements of the interfacial stress were carried out on the zirconia crowns.

The crown samples were wet sectioned diagonally into two halves using a low speed diamond saw as shown in Figure 1. No further surface preparation was done to limit further disturbing the stress state. The half sections were then mounted onto aluminum mounting tabs using a high strength rapid curing Bisphenol-A epoxy (Loctite Quickset 1 Min). The mounting tabs were attached to high precision kinematic bases which facilitate quick and precise alignment on the diffraction stage. Since the tabs were fixed to the base with a set screw, this mechanism provided freedom in rotating the sample about the surface normal.

Synchrotron setup

The residual stress measurements were conducted on the beamline 12.3.2 at the Advanced Light Source, Berkeley CA. The focusing optics consisting of elliptically bent Kirkpatrick-Baez mirrors were capable of achieving a beam cross section of 2 μm x 10 μm. Monochromatic X-ray beams with an energy tuned between 5 keV and 14 keV can be achieved. An X-ray energy of 7 keV was chosen to perform the experiment on the zirconia core crown in a 90° reflection geometry [17]. At higher energy, the diffraction cones shift towards the lower 2θ, and are spaced closely leading to a reduced resolution per ring. The CCD had 1024 x 1024 pixels and a resolution of 160 μm/pixel at a sample-to-camera distance of 78.9 mm. Prior to conducting the experiment, accurate calibration of distance and camera tilts was done by collecting a polychromatic diffraction pattern from a single crystal silicon placed on the sample. The intensity flux of the beam and energy was tuned by collecting a monochromatic diffraction pattern also from the single crystal silicon. It is critical to align the diffraction surface at the focal point where beam divergence is minimal, to reduce the potential errors introduced by sample displacement. This step was carried out using laser alignment[18].

Figure 2 is a schematic of the orientation of the sample and the camera with respect to the incoming beam. The diffracting cones corresponding to the lower and higher 2 angles are also indicated. The diffraction rings incident on the CCD camera were the intersecting arcs of the diffraction cone with the plane of the CCD. The sample rotation along was fixed for a particular location of the sample.

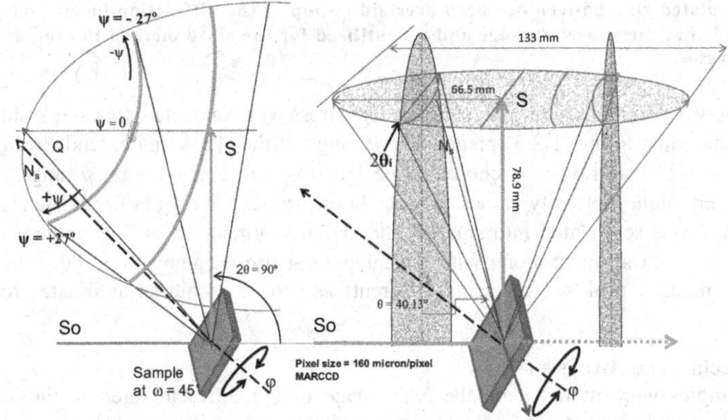

Figure 2 Schematic of the sample orientation and position of the CCD with respect to the diffraction cones. ψ is the angle subtended by the diffraction rings on the CCD. for a sample-to-CCD distance of 78.9 mm the maximum angle subtended was + 27° to - 27°.

Crystallographic information of zirconia

The zirconia core was made of yttria stabilized tetragonal polycrystalline zirconia with grain sizes on the order of 0.1 – 0.3 μm. The tetragonal poly-crystal structure can undergo stress induced phase transformation and form a low symmetry monoclinic crystal, with a larger lattice volume[19]. Figure 3 indicates a simulated fit based on a powder diffraction file (PDF) of a zirconia tetragonal crystal[20]. The PDF file contains information pertaining to the *d*-spacing of the different diffracting planes along with the corresponding relative intensity.

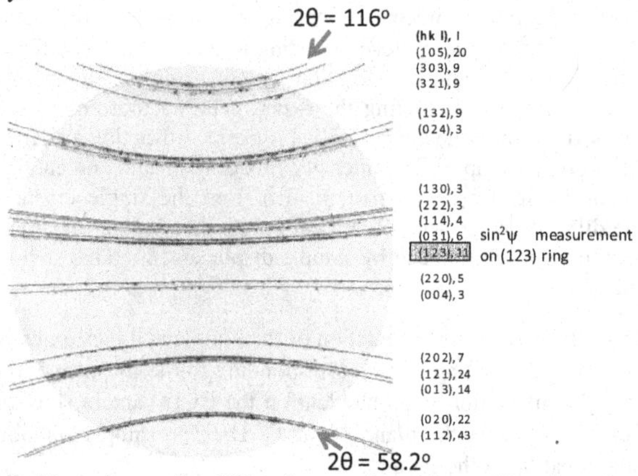

Figure 3 Monochromatic diffraction pattern from the zirconia core. A simulated ring pattern has been overlaid on top of the diffraction image. The (123) has the widest ψ range and was utilized for the $\sin^2\psi$ method to evaluate stresses.

The $\sin^2\psi$ method to measure the residual stress was conducted on a single diffraction ring corresponding to the (1 2 3) plane. The ψ range of the (123) ring extends from +27° to -27°. This particular ring was chosen since the ring had a maximum ψ range and also comparatively higher intensity values. Figure 4a. illustrates the range of ψ of the (123) ring. The ψ range was segmented into smaller integration segments of a 2° ψ range. Figure 4b indicates one such segment along with the integration profile along the 2θ direction. The integrated intensity peak was fit using a Lorentzian profile within an automated routine in XMAS[1]

Data Collection and Analysis

Samples were mounted on the XYZ stage of the diffractometer on the kinematic bases. The preliminary step of adjusting the surface height with respect to the focal point was carried out. The region of interest was mapped by acquiring exposures at a pre-defined exposure time after incremental translation of the sample. The images collected were in the MAR CCD format, and analyzed in XMAS. Based on the crystal information, the rings can be simulated and overlaid onto the experimentally obtained diffraction rings to confirm the calibration parameters [21].

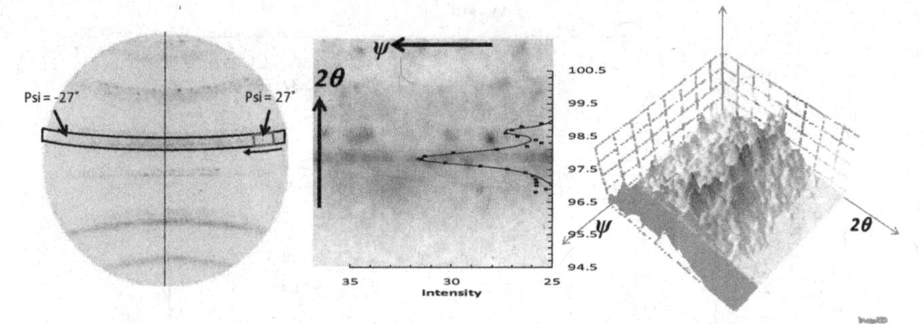

Figure 4 a. Raw diffraction image with the (123) ring highlighted and the total range of ψ it incorporates. b. Integrated intensity profile for a 2° segment of the (123) ring and the orientation of 2θ and ψ is indicated. c. Illustration of the angles in a 3D perspective indicates the signal-to-noise ratio at the given ψ segment.

XMAS contains an automated routine which can evaluate the *d*-spacing based on the peak fit for every image in a line or area scan. The precursor to the automation is a peak fit template. A template for the automation was prepared based on an initial peak fit on an exposure taken from a far field. Far from the interface, the diffraction rings were less distorted and had a better signal-to-noise ratio. The template provides an initial guess of the 2 value for the peak fitting program along with the tolerance and range of 2θ where the peak profile would be fit. The automated routine outputs the results in a delimited text format consisting of the 2θ value based on the peak fit along with the peak width, the integrated area under the peak, and the corresponding *d*-spacing and image index. Each automated output file belongs to a specific ψ segment. Starting with a peak fit template creation, the procedure was repeated for multiple ψ segments between the +27° to -27° ψ range. In the current analysis the (1 2 3) ring was divided into 27 segments with a 2° angular range. A customized VBA code was utilized in consolidating the results of *d*-spacing from the 37, ψ positions along the (123) ring. The residual stress for every scan point was evaluated using the $\sin^2\psi$ method. An example plot of the $\sin^2\psi$ versus the *d* spacing is shown in Figure 5. The slope of the line fit for the data is used. The intercept indicates the value of $d_{=0}$. Based on the slope and the intercept the residuals stress along a particular ϕ direction was determined according to the equation shown below.

$$\sigma_\phi = \frac{Slope}{d_{\psi=0}} \cdot \frac{E}{1+\nu} \qquad \text{Eqn. (1)}$$

RESULTS

The $\sin^2\psi$ method as explained in the previous sections was employed to obtain grain averaged residual stress. Monochromatic diffraction experiments were conducted on a zirconia core sample and a sectioned intact crown sample. Results were obtained from the analysis of the diffraction images acquired at line or area scans on the crowns. A typical *d* vs. $\sin^2\psi$ plot obtained through the automated analysis is shown in Figure 5.

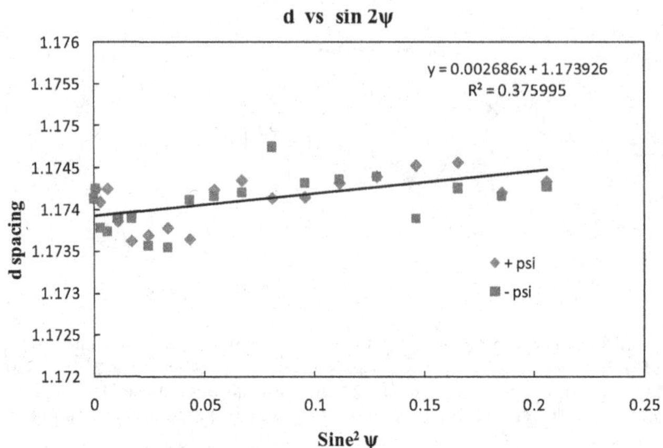

Figure 5 A typical *d*-spacing versus $\sin^2\psi$ plot obtained at a scan point on the zirconia core. Error bars in terms of stress are derived by calculating the standard deviation along the fit. The intercept of the line fit indicates the $d_{=0}$. Non-linearity of the *d* values from the positive and negative values typically indicates the presence of out-of-plane residual stress.

Zirconia core sample

Diffraction measurements on the zirconia core prior to veneering were used to validate the initial zero stress state. A line scan consisting of 15 scan points was performed on one of the cusps of the core. The location of the line scan was decided based on minimal curvature of the core since curvature leads to out of focus surfaces that increase the error bars of the measurement. Figure 6 illustrates the residual stress plot for a 150 μm line scan. An average biaxial residual stress of 10 MPa was determined on the core. A ± 5 μm displacement of the sample surface away from the focal point of the beam was observed for a 0.75 mm line scan on the sample surface. Considering the elastic modulus of zirconia (E = 205 GPa), the 10 μm total sample displacement represents a normal stress of 15 MPa. The error bars in stress were determined based on the error in the peak fit, which corresponds to a 2 range that in turn corresponds to a range in *d*-spacing. The associated strains were determined and converted to stress values.

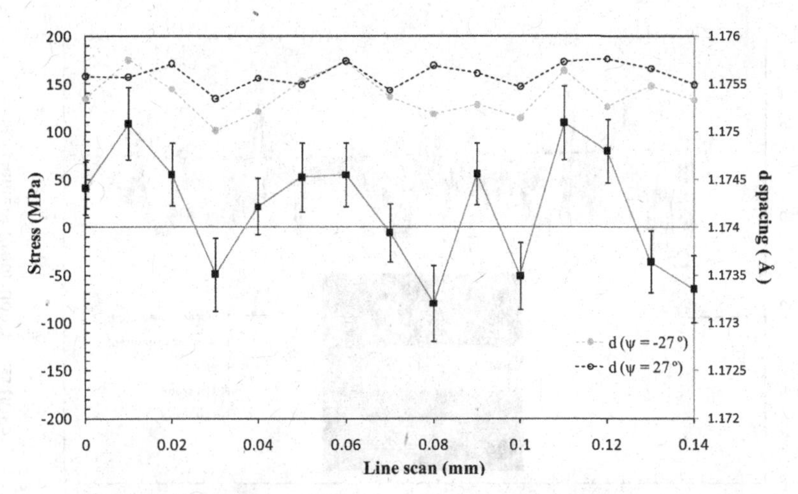

Figure 6 Biaxial stress measurements carried out on a zirconia core on one of the top cusps (indicated in the graphic). The line scan shows that the average type I residual stress observed on the core was close to zero. The line scan was recorded for 150 μm on the surface with curvature effects < 3 μm. If uncorrected, curvature introduces significant displacement errors in recorded values.

Sectioned Intact Crown

High resolution line scans with a 2 μm spatial resolution were performed at two different locations on a sectioned crown. Figure 7 illustrates results for the two line scans L_1 and L_2 as indicated in the picture set in the graph. L_1 was located towards the basal portion of the crown and L_2 was located around the top cusp, adjacent to the curved portion of the core. The average biaxial residual stress was found to be 257 MPa and 252 MPa for lines L_1 and L_2 respectively. An important point to note is shift in the stress profile towards compression in the 25 μm zone from the interface of the veneer and core. The evidence of such a trend was seen at both the line scans.

The corresponding values of *d*-spacing in the case of L_2 are shown in Figure 8. The plot exhibits a trend similar to the stress profiles indicating compression of the lattice planes in the vicinity of the interface. Finite element predictions were obtained from a model developed from the micro-tomography of the sample [22]. Predictions from the FE model indicate a compressive zone of approximately 225 MPa having a compressive stress on the porcelain side.

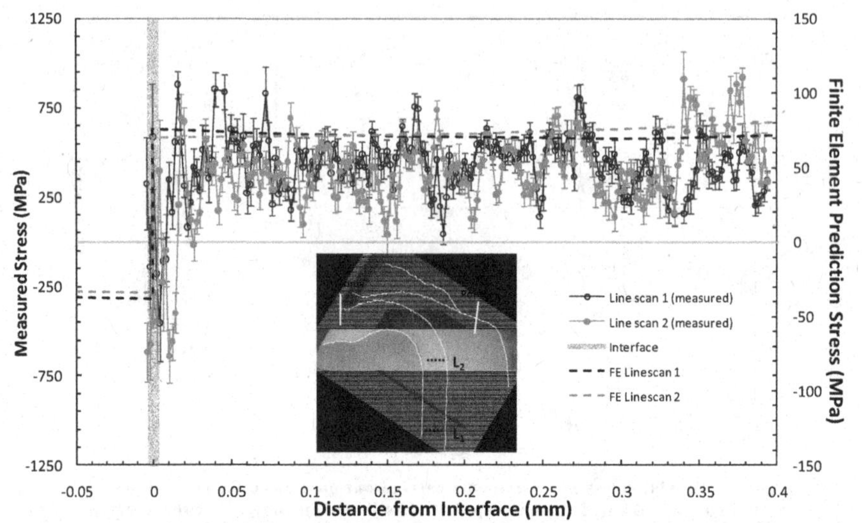

Figure 7 High resolution line scans L_1 & L_2 conducted on a sectioned intact crown. The line scans were located as shown in the micrograph. An average residual stress value of 325 MPa was observed for both line scans. Scatter in stresses (mostly between 100 – 400 MPa) can be seen along the line scan. At the interface, the stress values obtained from the tetragonal phase appear compressive up to about 300 MPa in both line scans. Results from finite element simulations of a real geometry of the crown are also overlaid as dotted lines corresponding to L_1 & L_2.

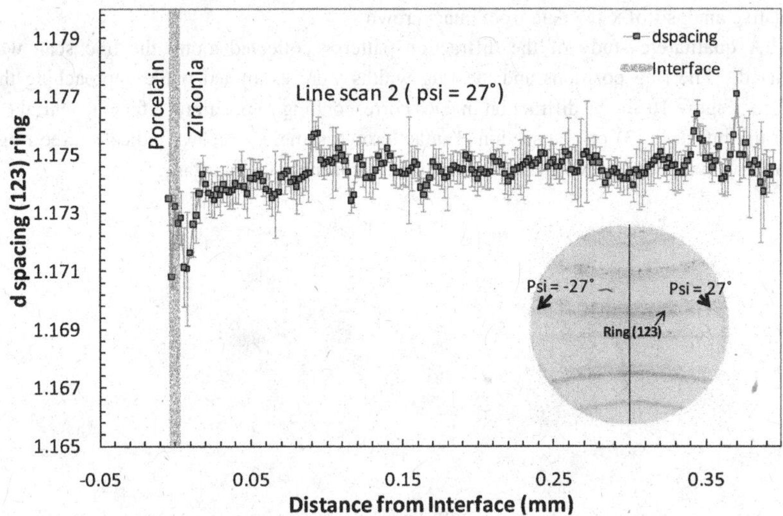

Figure 8 Illustration of the *d*-spacing variation at scan points along line scan L_2. The *d*-spacing of the (123) ring indicates compression at the interface.

Considering one of the exposures from a line scan at the interface of porcelain and zirconia, a grain contributing to the (123) diffraction cone was examined. The grain with largest deviation from the reference *d*-value was chosen to estimate the highest stress observed. For the grain observed in the inset in Figure 9, the measured strain based on the 2θ value was 3.658×10^{-3}, which corresponds to a stress of 750 MPa. Such analysis has not been previously reported.

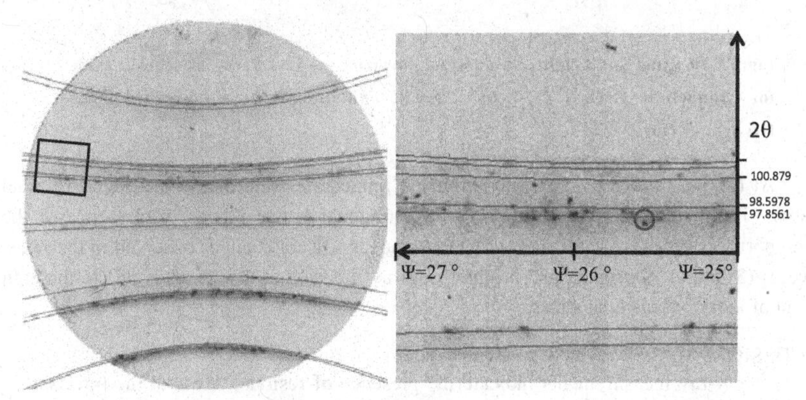

Figure 9 Illustration of the contribution of several grains to the (123) ring. Among many grains, the inset indicates a grain (circled red) with a large strain (3×10^{-3}) at ψ=35.5°.

Qualitative analysis of x-ray data from intact crown

A qualitative study of the diffraction patterns collected along the line scan was conducted. The ring positions and the ring widths were examined while approaching the interface. Figure 10 shows diffraction images corresponding to positions – far (1), central (2) and at the interface (3) on the sectioned intact crown sample. Arrows indicate three rings which show evidence of severe lattice distortion approaching the interface.

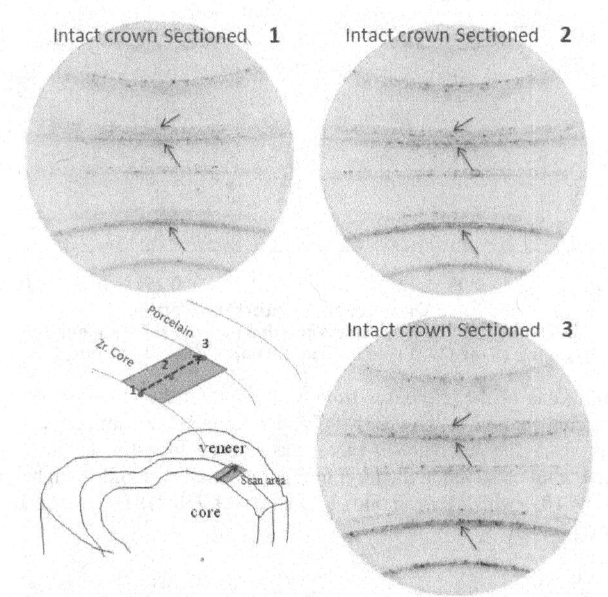

Figure 10 Ring gets brighter and wider towards the interface. As seen in table for Monoclinic ZrO_2 the (-3 0 2), (202), and the (013) rings overlap the tetragonal rings.

At (1) the grains are far away from the immediate zone of the interfacial residual stresses. Distinct rings corresponding to the tetragonal lattice planes were seen. At the central portion of the cross section (2) there was no significant changes observed in the rings. However (3) indicates broader and brighter rings as expected at the interface due to the high amount of localized residual stress.

DISCUSSION

The X-ray measurements indicate the presence of residual stress at the interface of porcelain and zirconia. In the case of the zirconia core without a veneer, negligible type I residual stress was observed. While expected, this baseline helps confirm the observed residual stresses were from the interface. Residual stress measurements using the $\sin^2$ method were performed only on the zirconia core region. Porcelain being amorphous, does not exhibit diffraction rings. A separate experimental technique would be required to assess the magnitude of residual stresses in porcelain [23]. Figure 6 indicates the residual stress in the monolithic zirconia core was close to zero. Use of a large beam compared to the zirconia grains, leads to a grain averaged residual stress value.

The process of firing the porcelain veneer on the zirconia core introduces interfacial residual stresses which is evident from Figure 7 - Figure 8. The addition of the veneer layer imposes an average tensile stress state of 250 MPa on the zirconia core as seen in Figure 7. Very high residual stresses are also evident locally as indicated in Figure 9. These stresses validate earlier polychromatic measurements [24] of similar magnitude. The sectioned samples were observed under an SEM which reveals rough bands resulting from the wet sectioning on a diamond saw (Shown in Figure 1). The striations from the sectioning cause a surface roughness with a maximum deviation from the surface of $\pm$ 5 μm which corresponds to a displacement error of up to 15 MPa in stress. Although the process of sectioning the sample would reduce or eliminate the existing out of plane stress, the $\sin^2\psi$ measurement successfully measures the biaxial in-plane stress. An evidence of the existence of some degree of out of plane residual stress can be observed in the nonlinearity of the *d*-spacing in Figure 5.

Compressive residual stresses of 350 $\pm$ 200 MPa can be observed in the zirconia at interface region (15 $\pm$ 2 μm of zirconia). At the interface where the residual stresses are expected to be higher, zirconia experiencing highest tensile stresses and porcelain under highest compressive stress, the tetragonal crystals of zirconia undergo a stress induced phase transformation. This transformation results in a monoclinic phase of zirconia with larger volume compared to the tetragonal crystal. Neighboring tetragonal zirconia crystals in the vicinity of the monoclinic transformed crystals are now subjected to a compressive stress due to the confinement within the bulk. Large error bars on the data points close to interface are due to the low intensity at higher ψ angles and broadened width of the peaks in the exposures.

A spotty wide inconsistent ring as seen in Figure 9 is indicative of the presence of excessive elastic deformation in some grains. Furthermore, if the ring is broad along ψ, the condition may be indicative of an overlapping secondary crystalline phase occurring due to transformations. Evidence of the monoclinic phase of zirconia close to the interface (20 μm from the interface) was investigated by examining individual rings in a mosaic pattern along the line scan. The mosaic of the inset depicts the rings corresponding to the (0 1 3) and (2 0 2) planes of the tetragonal phase of zirconia. As the interface was approached, these initial distinct appearing rings far from interface broaden and merge into each other. Furthermore, similar occurrence of 5 different low intensity peaks is seen overlapping the tetragonal peaks between the 2θ range of 96° to 99° (also seen in the inset mosaic in Figure 11).

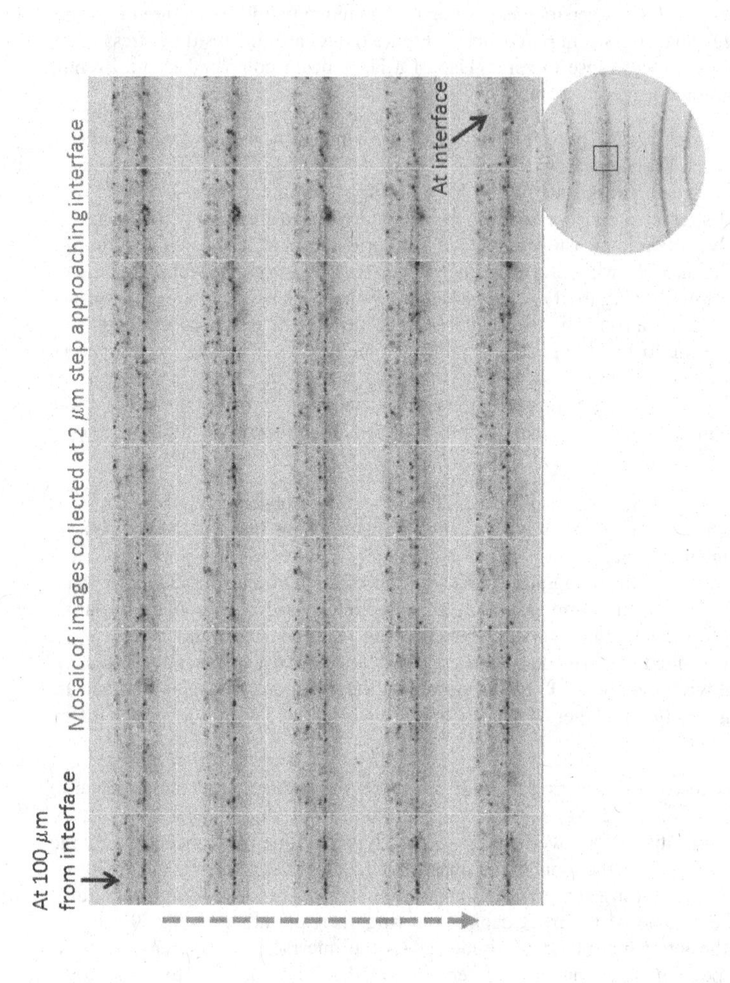

Figure 11 Mosaic pattern of the $\psi = 0$ segment for the (013) ring. The mosaic represents the rings obtained at 2 μm positions along the line scan starting from 100 μm away from the interface. The last few images correspond to the interface where the stress gradients were severe and caused distortion in the rings

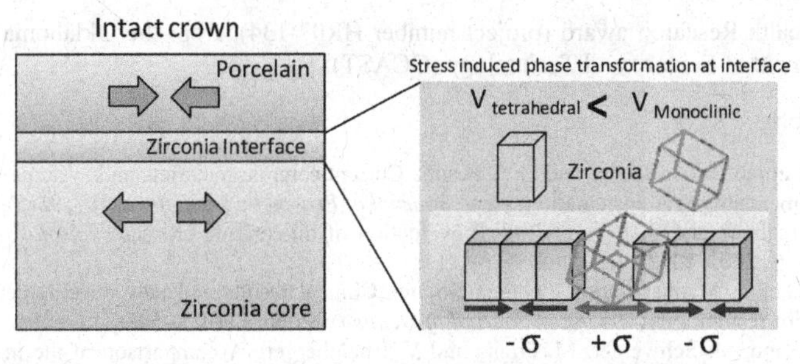

Figure 12 Illustration of the rationale of the observation of compressive stresses in zirconia at the interface layer due to stress induced phase transformations.

Figure 12 is a schematic of the state of the interface region of zirconia from an intact crown. Due to the mismatch in coefficient of thermal expansion of the porcelain and zirconia, the core is in a net tensile residual stress and porcelain in a net compressive residual stress. At the interface, the tetragonal crystals under high stress undergo stress induced phase transformation to form the monoclinic phase of zirconia. Transformation from the tetragonal to monoclinic phase results in a volume expansion locally. As shown in the schematic, this local volume expansion induces compressive stresses on the neighboring tetragonal zirconia crystals (indicated by blue arrows). This effect is manifested in the compressive trend seen earlier in line scans at the interface (Figure 7). From the analysis of the diffraction patterns it is estimated that the fraction of transformed monoclinic phase is small, less than 5%. A method to accurately quantify the transformed phase is desired.[25] Use of the micro-beam diffraction measurement is typically advantageous in such cases where the phases contributing to the stress are fractional.

CONCLUSIONS

A method for measuring grain averaged biaxial residual stress at a high resolution has been developed. Residual stresses were determined without the need of tilting the sample and without multiple exposures at a given location. High resolution line and area scans (2 μm spatial resolution) incorporating large numbers of data-points were performed. An automated data analysis routine was developed to compute biaxial residual stress values from 2D X-ray exposures.

Results from the experiments indicate the presence of a 250 MPa average tensile residual stress in the zirconia core. At the interface of intact crowns, the tetragonal zirconia crystals experience a compressive stress (350 ± 200 MPa) exerted by the phase transformed monoclinic zirconia crystals. The presence of monoclinic phase zirconia was confirmed through 2D qualitative analysis of the diffraction patterns.

ACKNOWLEDGEMENTS

We gratefully acknowledge the assistance of several researchers at NYU Dental College for discussion and providing samples including Dr. P. Coelho, Dr. V. Thompson, Dr. E. D. Rekow, M. Cabrera; and also M. Kunz at the Advanced Light Source for assistance with the experimental setup. Use of the Advanced Light Source is supported by the Director, Office of Science, Office of Basic Energy Sciences, of the U.S. Department of Energy under Contract No. DE-AC02-05CH11231. The research results discussed here were made possible in part by the

Oklahoma Health Research award (project number HR07-134), from the Oklahoma Center for the Advancement of Science and Technology (OCAST).

REFERENCES

1. H. J. Conrad, W. J. Seong, and G. J. Pesun: 'Current ceramic materials and systems with clinical recommendations: A systematic review', *Journal of Prosthetic Dentistry*, 2007, **98**(5), 389-404.
2. D. Gemalmaz and S. Ergin: 'Clinical evaluation of all-ceramic crowns', *Journal of Prosthetic Dentistry*, 2002, **87**(2), 189-196.
3. R. Hickel, J. Manhart, and F. Garcia-Godoy: 'Clinical results and new developments of direct posterior restorations', *American Journal of Dentistry*, 2000, **13**, 41D-54D.
4. W. Holand, M. Schweiger, M. Frank, and V. Rheinberger: 'A comparison of the microstructure and properties of the IPS Empress (R) 2 and the IFS Empress (R) glass-ceramics', *Journal of Biomedical Materials Research*, 2000, **53**(4), 297-303.
5. S. O. Hondrum: 'Review of the strength properties of dental ceramics', *Journal of Prosthetic Dentistry*, 1992, **67**(6), 859-865.
6. J. R. Kelly: 'Ceramics in restorative and prosthetic dentistry', *Annual Review of Materials Science*, 1997, **27**, 443-468.
7. S. C. Oh, J. K. Dong, H. Luthy, and P. Scharer: 'Strength and microstructure of IPS Empress 2 class-ceramic after different treatments', *International Journal of Prosthodontics*, 2000, **13**(6), 468-472.
8. G. Eskitascioglu, A. Usumez, M. Sevimay, E. Soykan, and E. Unsal: 'The influence of occlusal loading location on stresses transferred to implant-supported prostheses and supporting bone: A three-dimensional finite element study', *Journal of Prosthetic Dentistry*, 2004, **91**(2), 144-150.
9. J. Fischer, B. Stawarczyk, M. Tomic, J. R. Strub, and C. H. F. Hammerle: 'Effect of thermal misfit between different veneering ceramics and zirconia frameworks on in vitro fracture load of single crowns', *Dental Materials Journal*, 2007, **26**(6), 766-772.
10. S. K. Ganapathi, F. E. Spada, and F. E. Talke: 'Residual stresses and their effect on wear behavior of polycrystalline zirconia overcoated disks', *Magnetics, IEEE Transactions on*, 1992, **28**(5), 2533-2535.
11. J. C. Hanan and H. A. Bale: 'Analysis of interface stress, impact and damage in ceramic dental restoration materials', International Conference on Advanced Ceramics and Composites, Daytona, Florida, January, 2008, American Ceramics Society.
12. I. C. Noyan and J. B. Cohen: 'Residual Stress Measurement by Diffraction and Interpretation', ((Springer, New York, 1987)).
13. B. B. He and K. L. Smith: 'Fundamental equation of strain and stress measurement using 2D detectors.', *Proceedings of the Society of Experimental Mechanics Spring Conference on Experimental Mechanics, Society of Experimental Mechanics, Houston*, 1998, 217–220.
14. U. P. B B He, K L Smith: 'Advantages of using 2D detectors for residual stress measurements', *Advances in X-ray Analysis*, 2000, **42**, 429.
15. J. C. Middleton: 'Residual stresses and X-rays', *NDT International*, 1987, **20**(5), 291-294.
16. P. G. Coelho, E. A. Bonfante, N. R. F. Silva, E. D. Rekow, and V. P. Thompson: 'Laboratory Simulation of Y-TZP All-ceramic Crown Clinical Failures', *Journal of Dental Research*, 2009, **88**(4), 382-386.
17. N. Tamura, H. A. Padmore, and J. R. Patel: 'High spatial resolution stress measurements using synchrotron based scanning X-ray microdiffraction with white or monochromatic beam', *Materials Science and Engineering: A*, 2005, **399**(1-2), 92-98.
18. H. A. Bale, J. C. Hanan, N. Tamura, M. Kunz, P. G. Coelho, and V. Thompson: 'Interface Residual Stresses in Dental Zirconia Using Laue Micro-Diffraction.', *The Eighth International Conference on Residual Stresses*, 2008.
19. C. Piconi and G. Maccauro: 'Zirconia as a ceramic biomaterial', *Biomaterials*, 1999, **20**(1), 1-25.

20. E. H. Kisi and C. J. Howard: 'Elastic constants of tetragonal zirconia measured by a new powder diffraction technique', *Journal of the American Ceramic Society*, 1998, **81**(6), 1682-1684.
21. S. M. Polvino, N. Tamura, and O. Robach: 'X-MAS Manual v1', in 2005.
22. H. A. Bale, P. G. Coelho, and J. C. Hanan: '3D Finite Element Prediction of Thermal Residual Stresses in a Real Dental Restoration Geometry', 2010 (Unpublished work).
23. Y. Zhang and J. C. Hanan: 'Residual Stress in All-Ceramic Zirconia-Porcelain Crown Measured By Nanoindentation', 35th International Conference and Exposition on Advanced Ceramics and Composites, 2011.
24. H. A. Bale, N. Tamura, and J. C. Hanan: 'Micro-scale, localized high residual stresses in zirconia observed using micro x-ray diffraction', 2011 (In progress).
25. M. Allahkarami and J. C. Hanan: 'Characterization of non uniform veneer layer thickness distribution on curved substrate Zirconia ceramics using X-ray micro-tomography', 35th International Conference & Exposition on Advanced Ceramics & Composites, 2011.

---

[1] XMAS – X-ray Micro-diffraction Analysis Software developed by N. Tamura. (http://xraysweb.lbl.gov/microdif/XMAS_download.htm)

BIOACTIVE GLASS SCAFFOLDS FOR THE REPAIR OF LOAD-BEARING BONES

M.N. Rahaman, X. Liu, T.S. Huang

Department of Materials Sciences and Engineering, and Center for Bone and Tissue Repair and Regeneration, Missouri University of Science and Technology, Rolla, MO 65409

ABSTRACT

There is a need to develop scaffolds that can be used to repair defects in load-bearing bones, such as segmental defects in long bones, using a method that is biocompatible and durable during the patient's lifetime. Bioactive glass is an attractive scaffold material for use in filling bone defects because of its widely recognized ability to support the growth of bone cells and to bond firmly with hard and soft tissue. However, porous scaffolds of bioactive glass prepared by conventional methods often lack the requisite mechanical properties for repairing load-bearing bones. In this work, unidirectional freezing of suspensions and freeze extrusion fabrication (a solid freeform fabrication method) were used to create porous and strong scaffolds of silicate (13-93) bioactive glass. Oriented scaffolds with a columnar microstructure prepared by unidirectional freezing (porosity = 50%; average pore diameter = 100 μm) had a compressive strength of 35 ± 11 MPa and an elastic modulus of 7 ± 3 GPa, while scaffolds with a grid-like microstructure prepared by the freeze extrusion method (porosity = 50%; pore width = 300 μm) had compressive strength and elastic modulus values of 140 ± 70 MPa and 5.5 ± 0.5 GPa, respectively. The scaffolds supported the proliferation of osteogenic cells *in vitro*, as well as tissue infiltration *in vivo* when implanted in the calvaria of rats. These bioactive glass scaffolds are currently being evaluated for the repair of load-bearing defects in a rabbit tibia segmental defect model. Potential application of these bioactive glass scaffolds for the repair of loaded bone is discussed.

1. INTRODUCTION

There is a growing need for implants to repair bone defects caused by degenerative bone diseases, trauma, or congenital defects. In 2005 alone, there were 1.6 million bone graft procedures [1], and a significant fraction of these was used to treat bone defects in the elderly. As the population ages and life expectancy continues to rise, the number of bone graft procedures is expected to increase substantially. Autografts and allografts are commonly used to repair bone defects. These treatment methods have proved to be effective for the repair of contained defects in non-loaded bone which do not require a significant amount of graft material, but they suffer from limitations (e.g., donor site morbidity; limited supply; possible transmission of diseases; high costs). Furthermore, for many bone defects, the use of autografts and, to some extent, allografts, is not a viable treatment option. In particular, the repair and regeneration of large defects in load-bearing bones remains a challenging clinical problem. At present, bone allografts and custom metal augments are used to address segmental skeletal deficiency, but these treatments are limited by concerns related to high costs, limited availability, unpredictable long-term durability, uncertain healing to host bone, and other variables.

Porous synthetic scaffolds that replicate the structure and function of bone would be ideal bone substitutes, provided they have the requisite mechanical properties for reliable long-term cyclical loading during weight-bearing, far superior to those obtained with current methods [2]. The target mechanical properties of the scaffold are the subject of some debate, but a commonly-mentioned guideline is that the scaffold after *in vitro* tissue culture should exhibit mechanical properties which are comparable to those of the bone to be replaced. Additional characteristics of the scaffold which are considered to be essential for bone repair include biocompatibility (not toxic to cells and tissues) and the ability to support cell proliferation and differentiated function; bioactivity (ability to bond firmly to surrounding bone and soft tissues; osteoconductivity and osteoinductivity (ability to support new bone growth at the bone implant interface as well as within the interior of the scaffold); a porous three-dimensional (3D) architecture of interconnected pores to allow tissue infiltration, vascularization (formation of new blood vessels within the implant), and diffusion of nutrients into the interior of the scaffold; degradation into nontoxic products that can be easily resorbed or excreted by the body at a rate that matches the production of new bone; and the ability to be processed economically into anatomically relevant shapes and dimensions and be sterilized for clinical use.

Bioactive glass can provide many of the ideal properties of a scaffold material. Bioactive glass has a widely-recognized ability to convert to a hydroxyapatite (HA)-like material (the main mineral constituent of bone) and bond firmly with hard and soft tissues [3]. Bioactive glass is biocompatible, heals to host bone, degrades into non-toxic products, and can be fabricated into porous three-dimensional (3D) configurations. The degradation rate of the bioactive glass and its conversion to HA can be controlled by modifying the composition of the starting glass [4,5]. Porous scaffolds of bioactive glass are osteoconductive as well as osteoinductive. When implanted *in vivo*, bioactive glass releases ions (e.g., calcium and soluble silicon) that promote osteogenesis [6,7] and activate osteogenic gene expression [8,9].

A variety of methods can be used to form porous bioactive glass scaffolds. However, bioactive glass scaffolds fabricated using more conventional methods, such as sintering of particles or short fibers, gas foaming, or polymer foam replication, commonly have low strength (typically <20 MPa), so their use in the repair and regeneration of load-bearing bones is challenging. One approach for improving the mechanical properties is to use processing methods that provide greater control of the porosity and pore architecture of the scaffold. For example, unidirectional freezing of suspensions has been shown to create oriented bioceramic scaffolds with far higher mechanical properties (in the orientation direction) when compared to scaffolds with randomly-arranged pores [10]. However, oriented scaffolds prepared from aqueous suspensions typically have a lamellar microstructure in which the small pore widths (10–40 μm) are unfavorable for supporting tissue ingrowth. By modifying the composition of the solvent, such as the use of water dioxane mixtures [11], or the use of an organic phase (camphene) as the sublimable vehicle [12], oriented bioactive glass scaffolds have been prepared with pore diameters of 100 150 μm [13,14].

Solid freeform fabrication (SFF) or rapid prototyping methods provide the ability to create scaffolds with customized external shape and pre-designed internal architecture (porosity; pore size; and pore distribution) from computer generated models. It can improve current scaffold design by providing greater control of the internal architecture, dimensions, and anatomical shape of the scaffold [15]. Freeze extrusion fabrication (FEF) is a more recently developed SFF technique which has so far been used to fabricate dense ceramic articles for mechanical engineering applications [16]. In this technique, an aqueous mixture of ceramic particles and polymeric additives with a paste-like consistency is extruded through a nozzle to form filaments, which are frozen upon deposition to avoid slumping or distortion of the as-formed article. The rheology of the paste is critical for achieving the desirable plastic properties for flaw-free extrusion. Post processing steps in the formation of inorganic scaffolds by FEF consist of freeze-drying to sublime the frozen liquid, and heating to decompose the organic additives and then to thermally bond (sinter) the glass particles into a strong network.

The objective of this work was to investigate the use of two processing methods, unidirectional freezing of camphene-based suspensions and freeze extrusion fabrication, for the creation of porous and strong bioactive glass scaffolds. These methods were selected based on their potential for optimization of both the mechanical properties and the pore architecture of the scaffold. Silicate 13-93 bioactive glass was used because of its proven bioactivity [17] and our previous experience in the processing of this glass. Products of 13-93 bioactive glass are also approved for *in vivo* use by the U.S. Food and Drug Administration (FDA). The process variables for creating optimized bioactive glass scaffolds, the mechanical response and *in vitro* performance of the scaffolds, anf the ability of the scaffolds to support tissue infiltration *in vivo* were evaluated.

2. EXPERIMENTAL PROCEDURE

2.1 Materials

Melt-derived bioactive glass frits with the 13-93 composition (wt%): $53SiO_2$, $6Na_2O$, $12K_2O$, $5MgO$, $20CaO$, $4P_2O_5$) were kindly provided by Mo-Sci Corp., Rolla, Missouri. Particles of size finer than ~3 um were prepared by crushing and grinding the glass frits in a steel shatterbox (8500 Shatterbox®, Spex SamplePrep LLC., Metuchen, NJ, USA) followed by wet ball milling for 2 h in water with ZrO_2 grinding media. The dried glass particles were lightly ground to break up agglomerates, and used as the starting materials for fabricating porous scaffolds.

2.2 Unidirectional freezing of suspensions

The fabrication of porous scaffolds by unidirectional freezing of camphene-based suspensions consisted of a set of sequential steps: unidirectional freezing of the suspensions, thermal annealing of the frozen constructs near the softening point of the mixture to coarsen the camphene crystals, sublimation of the camphene, and heating to decompose the polymeric processing aids and to sinter the glass particles in the pore walls. The process is described in detail elsewhere [14]. Briefly, slurries consisting of 10 vol% glass particles, 2 wt% of isostearic acid (based on the dry mass of the particles) used as a dispersant, and liquid camphene were prepared by ball milling for 24 h at 55°C. Unidirectional freezing was performed by pouring the slurry into cylindrical Teflon molds (11 mm in diameter × 20 mm) placed on a copper plate kept at 3°C using an ice water mixture. After solidification, the samples were sealed with PVC caps to avoid camphene loss, and annealed for 24 h at 34°C. After the annealing step, the samples were cooled to room temperature, and the camphene was removed by sublimation (24 h at room temperature). The porous constructs were heated at 1°C/min to 500°C to burn out the isostearic acid, and then sintered for 1 h at 700°C (heating rate = 5 C/min) to densify the glass particles in the pore walls.

2.3 Freeze extrusion fabrication (FEF)

The FEF route consisted of extruding an aqueous mixture of glass particles and polymeric additives with a paste-like consistency through a nozzle, freezing the extruded filaments upon deposition to minimize distortion of the as-formed structure, sublimation of the ice, and heating to first burn out the polymeric additives and then to sinter the glass particles in the filaments (struts) [16]. The composition of the starting mixture used for preparing the paste is given in **Table I**. After ball milling the starting materials for 24 h to achieve a homogeneous mixture, a fixed mass of the liquid (10 wt%) was evaporated to give a paste-like consistency. The paste was vacuum degassed prior to loading into the FEF reservoir for extrusion.

The FEF system includes a computer-controlled gantry system (**Fig.1**) which is movable in three dimensions. The system is housed in an enclosure at −20°C. The paste is loaded into a reservoir consisting of a stainless steel sleeve and a plastic syringe, which are enclosed in a heating sleeve and kept at a fixed temperature (typically near room temperature) using a temperature controller, to prevent the paste from freezing in the syringe. The paste was extruded and deposited layer-by-layer on the working surface, with each layer deposited at 90° relative to the previous layer to form a grid-like pattern. Depending on the rheology of the paste, the software in the FEF machine was used to control the extrusion force and the rate of deposition. The nozzle diameter used in this work was 580 μm, and the distance between adjacent filaments was 600 μm. The as-formed FEF frozen constructs were dried in a freeze dryer (Model Genesis 25, Virits, Gardiner, NY) and then heated slowly (typically <1 C/min with a few isothermal holds) to burn out the polymeric additives, and sintered in air for 1 h at 700°C (heating rate = 5 C/min) to densify the glass struts.

Table I. Composition of starting slurry used in the preparation of the extrudable paste for FEF.

Component	Concentration (vol%)	Function	Manufacturer
13-93 glass particles	40.00	Solid phase	Mo-Sci Corp., Rolla, MO
EasySperse	0.50	Dispersant	ISP Technologies, Inc., Wayne, NJ
Surfnol	0.50	Defoamer	Air Products & Chemicals, Inc.,
Glycerol	1.00	WCCA*	Alfa Aesar, Ward Hill, MA
PEG 400	1.00	Lubricant	Alfa Aesar, Ward Hill, MA
Aquazol 5	4.00	Binder	ISP Technologies, Inc., Wayne, NJ
Deionized water	53.00	Solvent	—

* WCCA: Water crystallization control agent

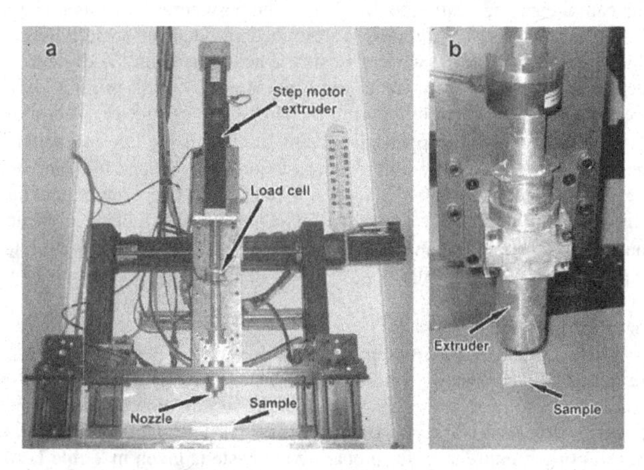

Fig.1. Photographs showing the components of the FEF equipment: (a) view of the gantry system and extruder; (b) magnified view showing the extruder.

2.4 Microstructural and mechanical characterization

The microstructure of the scaffolds and the glass phase was examined using optical microscopy (Optiphot-POL, Nikon Corp., Tokyo, Japan) and scanning electron microscopy, SEM (S-4700, Hitachi, Japan). The porosity of the constructs was determined from their weight and dimensions. The pore size distribution of the unidirectional scaffolds was determined using a liquid extrusion porosimeter (LEP-1100AX, Porous Materials Inc., NY). Synchrotron X-ray tomography was used to better characterize the three-dimensional pore network of the oriented scaffolds; scanning was carried out at the Advanced Light Source (ALS, Lawrence Berkeley National Lab., Berkeley, CA) using 22 keV monochromatic X-rays and a resolution size of 1.7 μm voxel.

The mechanical response of the scaffolds was measured in compression at a crosshead speed of 0.5 mm/min in an Instron testing machine (Model 4204; Instron, Norwood, MA). Samples were cut and ground using a diamond-coated wheel to produce a regular shape with flat contact surfaces. Samples prepared by unidirectional freezing had a cylindrical shape (7 mm in diameter 7 mm), while the samples prepared by FEF consisted of cubes (5 mm in length). Six samples were tested for each group of scaffolds, and the compressive strength and elastic modulus were determined as an average ± standard deviation.

2.5 *In vitro* cell culture

The ability of the bioactive glass scaffolds to support cell proliferation on the surface and into the interior pores was evaluated using the established MLO-A5 post-osteoblast/pre-osteocyte murine cell line. The cells were kindly provided by Dr. Lynda K. Bonewald, University of Missouri-Kansas City. The stock cells were maintained in collagen-coated plates (rat tail collagen type I, 0.15 mg/ml) containing α-MEM medium supplemented with 5% fetal bovine serum (FCS) and 5% new born calf serum (NCS) plus 100 U/ml penicillin. The dry-heat sterilized scaffolds were seeded with a cell suspension (1,000 cells/μl complete medium), incubated for 4 h to allow cell attachment, and then transferred to a 24-well plate containing 2 ml of complete medium per well. All cell cultures were maintained at 37°C in a humidified atmosphere of 5% CO_2, with the medium changed every 2 days.

After incubation for 2, 4, and 6 days, the scaffolds were removed, gently washed twice with warm phosphate-buffered saline (PBS), and placed in 2.5% glutaraldehyde in PBS to fix the cells. After an overnight soak in glutaraldehyde, the samples were washed with PBS and dehydrated thoroughly using a graded series of ethyl alcohol, followed by two soaks in hexamethyldisilazane (HMDS) for 10 min each. The samples were allowed to fully dry. The surfaces of the samples were sputter-coated with Au/Pd, and observed in the SEM (S-4700; Hitachi) at an accelerating voltage = 5 kV and a working distance = 12 mm. The surfaces and cross section of the scaffolds were also stained with a MTT assay to visualize the metabolically active cells on and within the scaffolds.

2.6 *In vivo* implantation

The ability of the scaffolds to support tissue ingrowth was tested using a non-healing rat calvaria defect model. This model was used because it is a standard assay to examine whether or not a material can generate new bone and heal a non-healing defect. Animal selection and management, surgical procedures, and preparation followed protocols approved by the Animal Care and Use Committee, Missouri University of Science and Technology. The animals (male Sprague Dawley rats) were anesthetized with a combination of ketamine (72 mg/kg) and xylazine (6 mg/kg), administered intraperitoneally. The surgical site was shaved and scrubbed with iodine. The periosteum was removed and a full-thickness bone defect (4.6 mm in diameter) was trephined in the center of each parietal bone using a saline-cooled trephine drill. The bone defects were implanted with disc-shaped bioactive glass scaffolds (4.6 mm in diameter × 1.5 mm), prepared by unidirectional freezing, in which the pores were oriented in the radial direction (**Fig. 2**).

Four weeks after implantation the animals were sacrificed by CO_2 inhalation. The specimens were fixed in 10% buffered formalin, dehydrated through a series of graded ethyl alcohol solutions, and embedded in poly(methyl methacrylate), PMMA. The sections were stained with Sanderson™ Bone Stain (Surgipath Medical Industries, Richmond, IL), and then counterstained with a solution consisting of 1 g acid fuchsin, 99 ml distilled water, and 1 ml acetic acid. This technique provided sufficient constrast to differentiate between soft tissue (blue), bone tissue (red), and osteoid production (purple) [18].

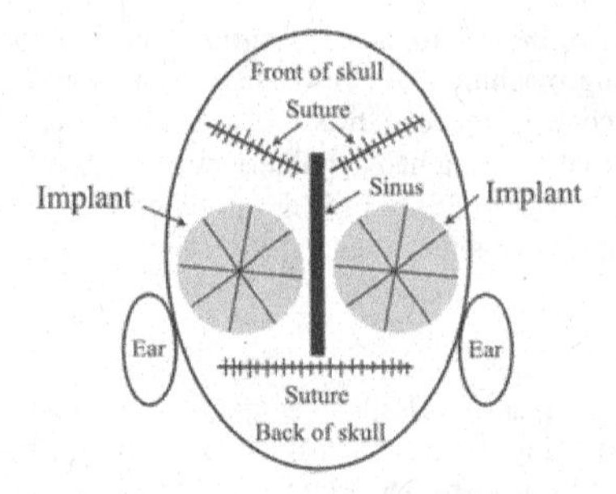

Fig. 2. Schematic showing the scaffold placement in the calvaria of a rat.

3. RESULTS

3.1 Microstructure of scaffolds

A sintered scaffold fabricated by the unidirectional freezing method is shown in **Fig. 3a**, while an SEM image of a cross section perpendicular to the freezing direction is shown in **Fig. 3b**. The glass phase was almost fully dense and the pores had an approximately circular cross section. Image analysis of cross sections perpendicular to the freezing direction showed that the scaffolds had a porosity of 50 ± 3%, made up of pores with an average size (diameter or width) of 100 μm. Synchrotron microCT (**Fig. 3c**) showed that the pores were oriented along the freezing direction. The pores were not isolated from each other, and each pore did not run directly from one end of the scaffold to the other. Instead, neighboring pores were connected at several positions along their length (**Fig. 3c**), indicating pore connectivity in the freezing direction as well as in the transverse direction. The pore size distribution measured using liquid extrusion porosimetry (**Fig. 4**) was in the range 50−150 μm with a peak at ~100 μm, which was in good agreement with the average pore size determined using image analysis.

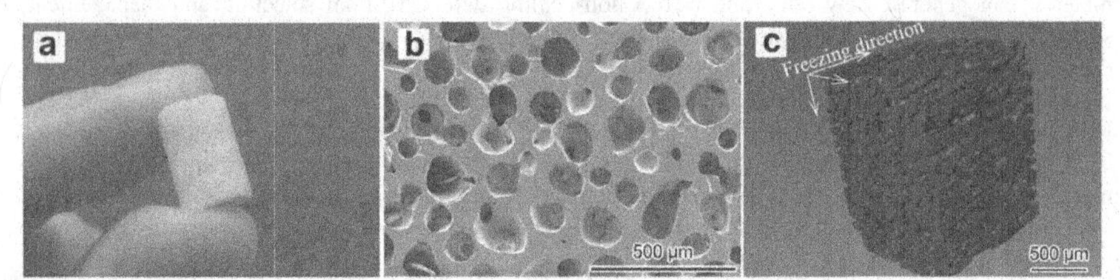

Fig. 3. (a) Optical image of a cylindrical scaffold (7 mm in diameter   10 mm) prepared by unidirectional freezing of suspensions; (b) SEM image of the cross section (perpendicular to the freezing direction) of the scaffolds; (c) X-ray microCT image showing the oriented pore phase.

Scaffolds fabricated using the FEF method had a pre-designed, grid-like microstructure (**Figs. 5a, 5b**). There was no observable distortion of the external shape of the construct from the forming step by FEF to the final sintering step. The glass filaments (struts), with a diameter (thickness) of   300 μm, were almost fully dense, and they appeared to be well bonded to the struts in adjacent layers (**Fig. 5c**). The pore width was ~300 μm in the plane of the deposition, and 150–200 μm in the direction perpendicular to the deposition plane.

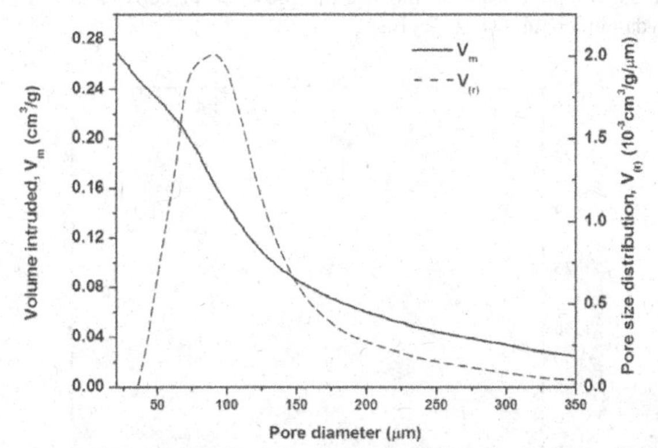

Fig. 4. Pore volume and pore size distribution of a sintered bioactive glass scaffold prepared by the unidirectional freezing method.

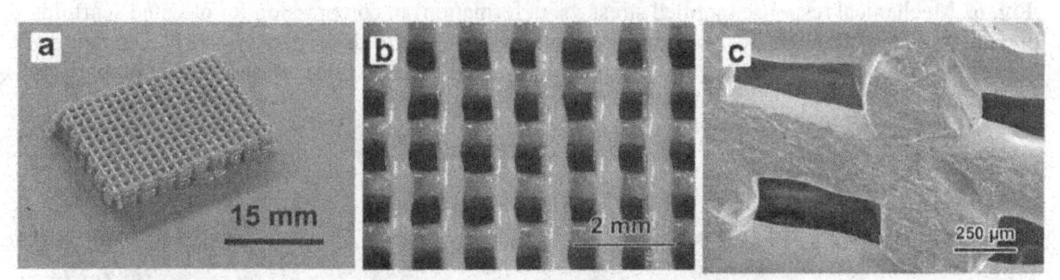

Fig. 5. (a) Optical image of a sintered 13-93 bioactive glass scaffolds (25 mm   15 mm   7 mm) prepared by FEF; (b) higher magnification image of the scaffold showing the uniform grid-like microstructure; (c) SEM image of a cross section parallel to the thickness direction of the scaffold.

3.2 Mechanical response

Both groups of scaffolds showed an elastic response in compression (**Fig. 6**). The stress increased approximately linearly with the deformation until the sample fractured into several pieces. This type of response is typically observed for dense ceramics and glasses. The compressive strength and elastic modulus of the oriented scaffold were 35 ± 11 MPa and 7.5 ± 2.5 GPa, respectively. The FEF scaffolds showed a compressive strength of 140 ± 70 MPa and elastic modulus of 5.5 ± 0.5 GPa.

3.3 *In vitro* response of scaffolds to cells

SEM images of oriented and FEF scaffolds seeded with MLO-A5 cells and incubated for 2, 4, and 6 days are showed in **Fig. 7**. The cell density increased with incubation time, showing the biocompatibility of the scaffolds and the ability of the scaffolds to support the proliferation of osteogenic cells. After incubation for 6 days, proliferation continued such that a layer of cells bridged the pores and almost completely covered the surface of the scaffold (**Figs. 7c, 7f**). The cells were observed to grow on the surface of the scaffolds and into the interior pores of both groups of scaffolds (**Figs. 7c, 7f**), despite of the difference in the pore size of the scaffolds,. MTT staining (purple) showed metabolically active cells on the surface and in the interior pores of the scaffolds (**Figs. 7c, 7f, inset**). A cross-section of the oriented scaffold showed the presence of cells deep into the interior of the scaffold after the 6-day incubation (**Fig. 7c, inset**).

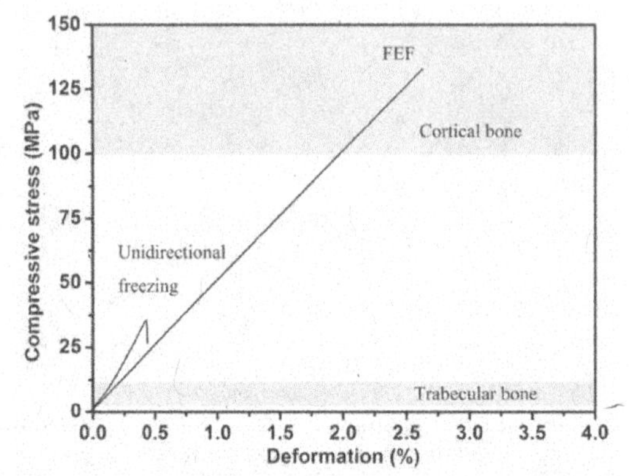

Fig. 6. Mechanical response (applied stress vs. deformation) in compression for oriented scaffolds prepared by unidirectional freezing of suspensions, and for scaffolds prepared by FEF. The range of values for the compressive strength of cortical bone and trabecular bone are shown for comparison.

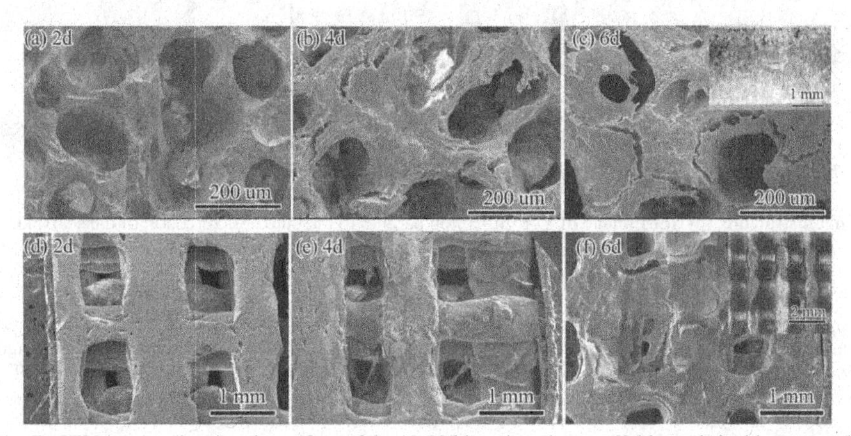

Fig. 7. SEM images showing the surface of the 13-93 bioactive glass scaffolds seeded with osteogenic MLO-A5 cells and incubated for 2, 4 and 6 days. (a)–(c): oriented scaffolds prepared by unidirectional freezing; (d)–(f): scaffolds with a grid-like microstructure prepared by FEF. (c) Inset: Freeze-fracture surface of oriented scaffold (incubation time = 6 days), showing MTT-labeled, metabolically-active cells (purple color) witnin the interior of scaffold; (f) Inset: Surface of FEF scaffold showing MTT-labeled cells on the surface of the scaffold (incubation time = 6 days).

3.4 Histology

Figure 8 shows a section of an oriented bioactive glass implant which was stained with Sanderson™ Bone Stain; the scaffold was implanted for 4 weeks in the calvaria of a rat. The scaffold retained its shape, and was well integrated with surrounding tissue. However, for this implantation time, only limited bone ingrowth (pink) occurred, mainly at the edge of the scaffold (**Fig. 8b**). The presence of osteocytes (arrow head) and osteoblasts (arrow) indicated that new bone formation was still in progress. Soft tissue (blue) infiltrated into the pores of the scaffold, and grew along the pore orientation direction as well as through the necks of the interconnected pores (**Fig. 8c**). These results showed that infiltration of new bone into the scaffold was limited after an incubation time of 4 weeks. Longer implantation times (12 and 24 weeks) are being used in ongoing experiments.

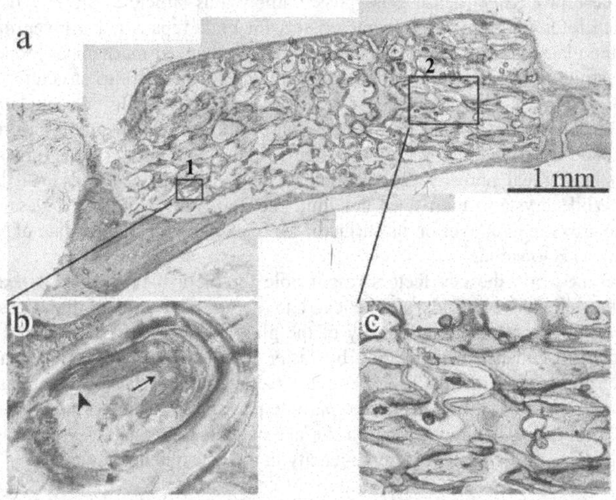

Fig.8. (a) Section of oriented bioactive glass implant stained with Sanderson™ Bone Stain. The scaffold was implanted for 4 weeks in a rat calvaria defect (b) Magnified image of boxed area 1, showing bone ingrowth (pink) into a pore near the edge of the scaffold, as well as osteocytes (arrow head) and osteoblasts (arrow); (c) magnified image of boxed area 2, showing soft tissue ingrowth (blue) into the interior pores.

4. DISCUSSION

4.1 Microstructure of bioactive glass scaffolds

Morphological features such as porosity, pore size, and pore interconnectivity have a critical effect on the ability of the scaffold to support tissue infiltration and on the mechanical properties of the scaffold. A porosity of 50% and interconnected pores with a neck diameter (or width) of 100 μm between neighboring pores are considered to be the minimum requirements to permit tissue ingrowth and function [19]. On the other hand, the mechanical properties (e.g., strength and elastic modulus) of the scaffold decrease with increasing porosity, and also depend on the size and orientation of the pores.

The creation of scaffolds with the ability to support tissue ingrowth and the requisite mechanical properties to substitute load-bearing bones requires adequate control of the microstructure.

The fabrication methods used in this work resulted in the creation of bioactive glass scaffolds with approximately the same porosity (~50%) but with two very different microstructures. Scaffolds prepared by the unidirectional freezing route had an oriented pore architecture, with a mean pore diameter (or width) of ~100 μm (**Figs. 3a, 4**). However, the pores were not perfectly aligned along the freezing direction; neighboring pores were connected at several positions along their length (**Fig. 3b**). Scaffolds prepared by FEF had a grid-like microstructure, with pores of width 300 × 300 μm in the plane of deposition and 150–200 μm in the thickness direction (**Fig. 5**).

For both groups of scaffolds, the glass phase making up the pore walls of the oriented scaffolds or the struts of the FEF scaffolds was almost fully dense, and the glass remained amorphous after the sintering step. These properties (dense glass phase; amorphous structure) are beneficial for creating bioactive glass scaffolds with optimum characteristics for bone repair. For a given microstructure, a high density of the glass phase is beneficial for achieving enhanced mechanical properties, since the glass walls or struts of the porous scaffold form the load-bearing structure. Avoidance of crystallization of the glass is beneficial for achieving a high density during sintering and for maintaining the bioactivity of the glass. Crystallization of a glass during sintering is known to reduce the sintering rate, and the formation of a high enough volume fraction of crystals (greater than ~15–20 vol%) can result in residual porosity [20], leading to glass ceramic scaffolds with low mechanical properties [21]. While crystallization does not limit the ability of a bioactive glass to be bioactive (formation of a hydroxyapatite layer on its surface), the bioactive potential (or rate of formation of the hydroxyapatite layer) is lowered.

In the present work, the key factors responsible for the achievement of a nearly fully dense glass phase in the sintered scaffolds are believed to be the fine size of the glass particles and, presumably, the homogeneous packing density of the glass particles in the as-formed scaffolds. Fine particles are known to enhance densification by faster viscous flow sintering of glass. In both the unidirectional freezing method and the FEF method, a stable colloidal suspension or paste was used for the creating the as-formed construct. While the particle packing in the as-formed constructs were not studied in detail, processing methods based on the use of colloidally stable systems are known to be beneficial for enhancing particle packing homogeneity in the consolidated system.

4.2 Mechanical properties

The present work shows promising mechanical properties for the potential use of these two groups of bioactive glass scaffolds in the repair of large defects in load-bearing bones, such as segmental defects in long bones. **Table II** shows the compressive strength and pore characteristics of bioactive glass scaffolds prepared by a variety of methods. This list is not meant to be comprehensive; instead, it shows representative properties of scaffolds prepared by each method. For comparison, **Table III** shows the porosity and mechanical properties (strength; elastic modulus) of human bone in compression. Bone is generally composed of two types: cortical (or compact) bone, and trabecular (cancellous or spongy) bone. Cortical bone, found primarily in the shaft of long bones and as the outer shell around cancellous bone, has a porosity of 5–10%; the compressive strength and elastic modulus in the direction parallel to the orientation (long axis) have been reported in the range 100–150 MPa and 10–20 GPa, respectively. A wide range has been reported for the elastic modulus (0.1–5 GPa) and the compressive strength (2–12 MPa) of cancellous bone [22, 23].

According to the data in Tables II and III, bioactive glass scaffolds prepared by methods such as sintering of particles or short fibers, polymer foam replication, and sol-gel processing have compressive strengths in the range of those for trabecular bone. The use of these scaffolds for repairing large defects in load-bearing bone is challenging. On the other hand, oriented bioactive glass scaffolds prepared in this work by unidirectional freezing of suspensions have an average compressive strength (35 MPa) that is approximately three times the highest strength (12 MPa) reported for trabecular bone. Furthermore, scaffolds prepared by FEF have an average compressive strength (140 MPa) in the range reported for cortical bone.

As indicated above, one factor that might contribute to the high compressive strength of the scaffolds created in this work is the high density of the glass walls of the oriented scaffolds or the glass struts in the FEF scaffolds. However, the oriented bioactive glass scaffolds have an average compressive strength (35 MPa) that is approximately four times lower than that for the FEF scaffold (140 MPa), even though the porosity of the scaffolds are approximately the same (50%). Examination using SEM and X-ray microCT indicated that the pores in the oriented scaffolds are not perfectly aligned, but that they are connected to adjacent pores at several points along their length. Our recent work is showing that with better alignment of the pores through controlling the freezing rate, far higher compressive strengths (50 MPa) can be obtained for scaffolds with porosity and pore diameter similar to those for the scaffolds prepared in this work.

Table II. Methods used to create silicate bioactive glass scaffolds, and characteristics of the fabricated scaffolds.

Method	Glass	Porosity (%)	Pore size (μm)	Strength* (MPa)	Reference
Thermal bonding of:					
Particles	13–93	40–45	100–300	22 ± 1	Fu et al. (2007) [24]
Short fibers	13–93	45–50	>100	5	Jung et al. (2010c) [25]
Polymer foam replication	45S5	89–92	510–720	0.4 ± 0.1	Chen et al. 2006 [21]
	13–93	75–85	100–500	11 ± 1	Fu et al. (2008c) [26]
Sol-gel foam	70S30C	82	500 (100)#	2.4	Jones et al (2006) [27]
Unidirectional freezing of suspensions	13–93	53–57	90–110	25 ± 3	Fu et al. (2010c) [10]
	13–93	50–55	60–120	27 ± 8	Liu et al. (2011a) [14]
	13–93	50	50–150	35 ± 11	This work
Rapid prototyping:					
Selective laser sintering	13–93	58–60	700– 1000	15 ± 1	Velez (2010) [28]
Freeze extrusion fabrication	13–93	50	100–500	140 ± 70	This work
Robocasting	6P53B	60	500	135 ± 20	Fu et al. (2010) [29]

*Compressive loading; #Macropore diameter = 500 μm; interconnected pore diameter = 100 μm.
Glass composition (wt%): 45S5: 45 SiO_2, 24.5 Na_2O, 24.5 CaO, 6 P_2O_5; 13-93: 53 SiO_2; 6Na_2O, 12 K_2O, 5 MgO, 20 CaO, 4 P_2O_5; 6P53: 52.6 SiO_2; 10.4 Na_2O, 2.8 K_2O, 10.2 MgO, 18 CaO, 6 P_2O_5; 70S30C: 71.4 SiO_2, 28.6 CaO

Table III. Summary of the porosity and mechanical properties of human bone in compression.

Type of bone	Porosity (%)	Strength (MPa)	Elastic modulus (GPa)
Cortical	5–10	100–150	10–20
Trabecular	30–90	2–12	0.1–5

4.3 Potential use of fabricated bioactive glass scaffolds in bone repair

In vitro, the two groups of scaffolds created in this work supported the proliferation of osteogenic MLO-A5 cells on the surface and into the interior pores of scaffold (**Fig. 7**). *In vivo*, previous work showed that scaffolds with pore characteristics similar to those of the oriented scaffolds prepared in this work supported tissue infiltration after implantation for 4 weeks in subcutaneous pockets in the dorsum of rats [30]. In this work, a rat calvaria defect model was used to evaluate the ability of the bioactive glass scaffolds to support new bone formation and to heal a non-loaded bone defect. The studies showed that the oriented bioactive glass scaffolds integrated well with surrounding bone after implantation 4 weeks, and supported soft tissue infiltration into the interior of the scaffolds. However, only limited new bone formation was observed, near the edge of the implant. It appeared that the implantation time was too short to support bone infiltration into the scaffolds (**Fig. 8**). *In vivo* experiments are in progress in which longer implantation times (12 and 24 weeks) are being used.

5. CONCLUSION

Two groups of 13-93 bioactive glass scaffolds, prepared by unidirectional freezing of suspensions and freeze extrusion fabrication (FEF), have the requisite pore characteristics known to be favorable for supporting tissue infiltration and adequate mechanical strength for the repair of load-bearing bones. Scaffolds with an oriented microstructure (porosity $\approx$50%, average pore diameter or width $\approx$100 μm) had compressive strength and elastic modulus values of 35 ± 11 MPa and 7 ± 3 GPa, respectively. Scaffolds prepared by FEF, with a grid-like microstructure (porosity $\approx$50%; pore width $\approx$300 μm), had compressive strength and elastic modulus values of 140 ± 70 MPa and 5.5 ± 0.5 GPa, respectively. Both groups of scaffolds supported the proliferation of osteogenic MLO-A5 cells on the surface and into the interior pores of the scaffold. Implantation of the oriented scaffolds for 4 weeks in rat calvaria defects showed good integration of the scaffolds with surrounding bone and infiltration of soft tissue into the interior pores, but new bone formation was limited to the edge of the scaffolds. Longer implantation times are required to evaluate new bone formation in the scaffolds.

ACKNOWLEDGEMENT

This work was supported by the National Institutes of Health (NIAMS), Grant # 1R15AR056119, and by the U.S. Army Medical Research Acquisition Activity, under Contract No. W81XWH-08-1-0765. The authors would like to thank Dr. D. E. Day, Dr. R. F. Brown, and Dr. M. C. Leu, Missouri S&T, for their collaboration, and Mo-Sci Corporation, Rolla, Missouri, providing the bioactive glass used in this work.

REFERENCES

1. U.S. Census Bureau. Health and Nutrition. U.S. Census Bureau Statistical Abstracts of the United States, Washington, DC, 117 (2009).
2. D.W. Hutmacher, Scaffold Design and Fabrication Technologies for Engineering Tissues state of the art and Future Perspectives, *J. Biomater. Sci. Polymer. Edn.*, 12, 107-24 (2001)
3. L.L Hench, Bioceramics, *J. Am. Ceram. Soc.*, 81, 1705-28 (1998).
4. W. Huang, D.E. Day, K. Kittiratanapiboon, M.N. Rahaman, Kinetics and Mechanisms of the Conversion of Silicate (45S5), Borate, and Borosilicate Glasses to Hydroxyapatite in Dilute Phosphate Solutions, *J. Mater. Sci.: Mater. Med.*, 17, 583-96 (2006).
5. A. Yao, D.P. Wang, W. Huang, M.N. Rahaman, D.E. Day, *In Vitro* Bioactive Characteristics of Borate-Based Glasses with Controllable Degradation Behavior, *J. Am. Ceram. Soc.*, 90, 303-6 (2007).

6. D.L. Wheeler, K.E. Stokes, H.E. Park, J.O. Hollinger, Evaluation of Particulate Bioglass® in a rabbit radius ostectomy model. *J. Biomed. Mater. Res.*, 35, 249-54 (1997).
7. D.L. Wheeler, K.E. Stokes, R.G. Hoellrich, D.L. Chamberland, S.W, McLoughlin, Effect of Bioactive Glass Particle Size on Osseous Regeneration of Cancellous Defects, *J. Biomed. Mater. Res.*, 41, 527-33 (1998).
8. I.D. Xynos, A.J. Edgar, L.D. Buttery, L.L. Hench, J.M. Polak, Ionic Products of Bioactive Glass Dissolution Increase Proliferation of Human Osteoblasts and Induce Insulin-like Growth Factor II mRNA Expression and Protein Synthesis, *Biochem. Biophys. Res. Commun.*, 276, 461-5 (2000).
9. I.D. Xynos, A.J. Edgar, L.D. Buttery, L.L. Hench, J.M. Polak, Gene-expression Profiling of Human Osteoblasts Following Treatment with the Ionic Products of Bioglass 45S5 Dissolution, *J. Biomed. Mater. Res.*, 55, 151-7 (2001).
10. Q. Fu, M.N. Rahaman, B.S. Bal, R.F. Brown, Preparation and *In Vitro* Evaluation of Bioactive Glass (13-93) Scaffolds with Oriented Microstructures for Repair and Regeneration of Load-bearing Bones, *J. Biomed. Mater. Res.*, 93A, 1380-90 (2010).
11. Q. Fu, M.N. Rahaman, F. Dogan, B.S. Bal, Freeze Casting of Porous Hydroxyapatite Scaffolds – I. Processing and General Microstructure, *J. Biomed. Mater. Res.*, 86B, 125-35 (2008).
12. K. Araki, J.W. Halloran, New Freeze-casting Technique for Ceramics with Sublimable Vehicles, *J Am Ceram Soc*, 87, 1859-63 (2004).
13. B.H. Yoon, W.Y. Choi, H.E. Kim, J.H. Kim, Y.H. Koh, Aligned Porous Alumina Ceramics with High Compressive Strengths for Bone Tissue Engineering. *Scripta Mater*, 58, 537-40 (2008).
14. X. Liu, M.N. Rahaman, Q. Fu, Oriented Bioactive Glass (13-93) Scaffolds with Controllable Pore Size by Unidirectional Freezing of Camphene-based Suspensions: Microstructure and Mechanical Response, *Acta Biomaterialia*, 7, 406-16 (2011).
15. E. Sachlos, J.T. Czernuszka, Making Tissue Engineering Scaffolds Work. Review: the Application of Solid Freeform Fabrication Technology to the Production of Tissue Engineering Scaffolds, *European Cells & Mater.*, 5, 29-39 (2003).
16. T.S. Huang, M.S. Mason, X.Y. Zhao, G.E. Hilmas, M.C. Leu, Aqueous-based Freeze-form Extrusion Fabrication of Alumina Components, *Rapid Prototyping J.*, 15, 88-95 (2009).
17. M. Brink, T. Turunen, R. Happonen, A. Yli-Urpo, Compositional Dependence of Bioactivity of Glasses in the System Na_2O-K_2O-MgO-CaO-B_2O_3-P_2O_5-SiO_2, *J. Biomed. Mater. Res.*, 37, 114-21 (1997).
18. M.L. Wang, J. Massie, R.T. Allen, Y.P. Lee, C.W. Kim. Altered Bioreactivity and Limited Osteoconductivity of Calcium Sulfate-based Bone Cements in the Osteoporotic Rat Spine. *Spine J.*, 8, 340-50 (2008).
19. S.F. Hulbert, F.A. Young, R.S. Mathews, J.J. Klawitter, C.D. Talbert, F.H. Stelling, Potential of Ceramic Materials as Permanently Implantable Skeletal Prostheses, *J. Biomed. Mater. Res.*, 4, 433-56 (1970).
20. M.N. Rahaman, Sintering of Ceramics, CRC Press, Boca Raton, FL, 2007, pp. 289-97.
21. Q.Z. Chen, I.D. Thompson, A.R. Boccaccini, 45S5 Bioglass®-derived Glass-Ceramic Scaffolds for Bone Tissue Engineering, *Biomaterials*, 27, 2414-25 (2006).
22. J.Y. Rho, M.C. Hobatho, R.B. Ashman, Relations of Density and CT Numbers to Mechanical Properties for Human Cortical and Cancellous Bone, *Med. Eng. Phys.*, 17, 347-55 (1995)
23. Y.H. An, R.A. Draughn. Mechanical Testing of Bone and the Bone-implant Interface. CRC Press, Boca Raton, FL, 2000; p. 51.
24. Q. Fu, M.N. Rahaman, W. Huang, D.E. Day, and B.S. Bal, Preparation and Bioactive Characteristics of a Porous 13-93 Glass, and its Fabrication into the Articulating Surface of a Proximal Tibia, *J. Biomed. Mater. Res.*, 82A, 222-9 (2007).

25. S. Jung, Borate Based Bioactive Glass Scaffolds for Hard and Soft Tissue Engineering, PhD Thesis, Missouri University of Science and Technology, 298, (2010).
26. Q. Fu, M.N. Rahaman, B.S. Bal, R.F. Brown, D.E. Day, Mechanical and *In Vitro* Performance of 13-93 Bioactive Glass Scaffolds Prepared by a Polymer Foam Replication Technique, *Acta Biomater.*, 4, 1854-64 (2008).
27. J.R. Jones, L.M. Ehrenfried, L.L. Hench, Optimising Bioactive Glass Scaffolds for Bone Tissue Engineering, *Biomaterials*, 27, 964-73 (2006).
28. M. Velez, K.C.R. Kolan, M.C. Leu, R.F. Brown, G.E. Hilmas, Selective Laser Sintering Fabrication of Bioglass Bone Scaffolds, Paper presented at the Materials Science & Technology Annual Meeting, Houston, TX, 2010.
29. Q. Fu, E. Saiz, A.P. Tomsia, Bio-inspired Highly Porous and Strong Glass Scaffolds, *Adv. Funct. Mater.,* In press (2010).
30. Q. Fu, M. N. Rahaman, B. S. Bal, K. Kuroki, R. F. Brown. *In Vivo* Evaluation of 13-93 Bioactive Glass Scaffolds with Trabecular and Oriented Microstructures in a Subcutaneous Rat Implantation Model. *J. Biomed. Mater. Res.*,95A, 235-44 (2010).

DO CELL CULTURE SOLUTIONS TRANSFORM BRUSHITE ($CaHPO_4 \cdot 2H_2O$) TO OCTACALCIUM PHOSPHATE ($Ca_8(HPO_4)_2(PO_4)_4 \cdot 5H_2O$)?

Ibrahim Mert, Selen Mandel, and A. Cuneyt Tas
Department of Biomedical Engineering, Yeditepe University, Istanbul 34755, Turkey

ABSTRACT

The purpose of this study was to investigate the transformation of brushite (dicalcium phosphate dihydrate, DCPD, $CaHPO_4 \cdot 2H_2O$) powders at 36.5 C in DMEM (Dulbecco's Modified Eagle Medium) solutions. Two sets of brushite powders with different particle shapes were synthesized to use in the above DMEM study. The first of these brushite powders was prepared by using a method which consisted of stirring calcite ($CaCO_3$) powders in a solution of ammonium dihydrogen phosphate ($NH_4H_2PO_4$) from 6 to 60 minutes at room temperature. These powders were found to consist of dumbbells of water lily-shaped crystals. The second one of the brushite powders had the common flat plate morphology. Both powders were separately tested in DMEM-immersion experiments. Monetite (DCPA, $CaHPO_4$) powders were synthesized with a unique water lily morphology by heating the water lily-shaped brushite crystals at 200 C for 2h. Brushite powders were found to transform into octacalcium phosphate (OCP, $Ca_8(HPO_4)_2(PO_4)_4 \cdot 5H_2O$) upon soaking in DMEM (Dulbecco's Modified Eagle Medium) solutions at 36.5 C over a period of 1 day to 1 week. Brushite powders were known to transform into apatite when immersed in synthetic (simulated) body fluid (SBF) solutions. This study found that DMEM solutions are able to convert brushite into OCP, instead of apatite.

INTRODUCTION

DMEM (Dulbecco's Modified Eagle Medium) solutions are used in cell culture as the growth medium. HEPES-buffered DMEM solutions contain inorganic salts (to supply Ca^{2+}, Mg^{2+}, Na^+, K^+, Fe^{3+}, $H_2PO_4^-$, HCO_3^- and Cl^- ions), amino acids, vitamins and glucose. DMEM solutions have a Ca/P molar ratio of 1.99.

SBF (simulated [1] or synthetic body fluid [2]) solutions, on the other hand, are usually TRIS- or HEPES-buffered [3] and only contain inorganic salts with an overall Ca/P molar ratio of 2.50.

DMEM solutions are one of the best media to test the bioactivity of synthetic biomaterials. Bioactivity of any material cannot be tested in SBF solutions. SBF solutions cannot be used in cell culture studies since they lack the necessary nutrients, such as amino acids, vitamins and glucose, to allow and sustain the proliferation of living organisms.

However, SBF solutions were heavily used in recent decades to test the so-called and deceptive bioactivity of a given material, regardless of being metallic, ceramic, glassy or polymeric. Can bioactivity be tested *in vitro* with SBF solution [4]? How useful is SBF in predicting *in vivo* bone bioactivity [5]? These two questions were previously asked and had actually become the exact titles of two articles cited here [4, 5]. SBF solutions are metastable and supersaturated solutions (they are supersaturated with respect to the formation of apatite-like, i.e., apatitic, calcium phosphate (CaP) phase) and SBF solutions would thus automatically precipitate apatitic CaP on substances with basic surfaces immersed in them. Precipitation of an apatite-like phase and bioactivity were simply two diverse concepts which were often confused in the literature very seriously [4, 5].

Nevertheless, the following question has not been asked frequently; could it be possible to test, *in vitro*, such a so-called bioactivity by using readily available DMEM solutions, containing amino acids, vitamins and glucose, instead of SBFs? The current study asks this question in search of an answer to it, by ageing brushite (dicalcium phosphate dihydrate, DCPD, $CaHPO_4 \cdot 2H_2O$) powders in DMEM solutions at 36.5 C.

Two different, chemically-synthesized $CaHPO_4 \cdot 2H_2O$ powders with quite different particle morphologies were tested by soaking them in DMEM solutions at the human body temperature of 36.5 C from 1 day to 1 week.

Brushite is a relatively high solubility (with log K_{sp} of -6.6 [6]) calcium phosphate compound and is known to convert into apatite-like calcium phosphate when soaked in SBF solutions at the human body temperature for about one week [7-21].

However, both of the brushite powders used in this study [22] were found to transform into octacalcium phosphate (OCP, $Ca_8(HPO_4)_2(PO_4)_4 \cdot 5H_2O$) when soaked in DMEM solutions. To the best of our knowledge, this study was the first to report on the bulk transformation of brushite powders into OCP upon soaking in a common cell culture solution, such as DMEM.

EXPERIMENTAL PROCEDURE

Synthesis of Brushite or Monetite with Water Lily-shaped (WL) Crystals

Brushite (DCPD, $CaHPO_4 \cdot 2H_2O$) powders with a unique water lily- morphology were synthesized as follows. 40.0 g (=0.3477 mol P) of ammonium dihydrogen phosphate, $NH_4H_2PO_4$ (≥99.9%, Cat. No: 1.01126, Merck KGaA, Darmstadt, Germany) was first dissolved in 340 mL of distilled water. This solution had a pH value of 3.9±0.1 at room temperature (RT). The solution was placed into a 500 mL-capacity Pyrex glass media bottle. 10.0 g (=0.0999 mol Ca) of calcite (precipitated-chalk type), $CaCO_3$ (≥99.9%, Cat. No: 12010, Riedel-de-Haen, Germany) powder was added into the bottle. The formed suspension was stirred (500 rpm) at RT for 30 minutes, by using a Teflon-coated magnetic stirrer. After 30 min of stirring, the pH value of the white suspension was measured to be 5.9±0.1. The particles of the suspension were recovered from their mother liquor by using a porcelain Buechner funnel containing a No. 3 Whatman filter paper. The funnel was attached to a mechanical vacuum pump during filtration. The wet cake on the filter paper was finally washed with 750 mL of distilled water. Obtained powders were dried in a clean watch-glass overnight at 75°C, in a static air microprocessor-controlled oven, to obtain 14.64±0.2 g of brushite with the water lily (WL, *Nymphaeaceae*) morphology. After the addition of precipitated-chalk type $CaCO_3$ powder into the above-mentioned $NH_4H_2PO_4$ solution, stirring was not a necessity; the same results would have been obtained if one were to choose to keep the resultant white suspension "non-stirred" for an overnight period (typically 18 to 19 hours).

The above was the optimized and up-scaled synthesis recipe of the WL-shaped brushite powders. Prior to the development of this recipe, the influence of stirring time on the advance of reaction was studied over the range of 90 seconds to 60 minutes. In these experiments, 10.0 g of $NH_4H_2PO_4$ was first dissolved in 85 mL of distilled water and then stirred at RT with 2.5 g of calcite ($CaCO_3$) powder in 100 mL-capacity glass bottles containing 85 mL distilled water.

To convert the WL-shaped brushite crystals into monetite (DCPA, $CaHPO_4$), 1.0 g of the above brushite powders was kept (in clean watch glasses) for 2 hours in a microprocessor-controlled static air oven pre-heated to 200°C. The following reaction was expected to take place during this heating;

$$CaHPO_4 \cdot 2H_2O\,(s) \rightarrow CaHPO_4\,(s) + 2H_2O\,(g) \qquad (1).$$

Synthesis of Brushite with Flat Plate-shaped (FP) Crystals

The synthesis procedure used to form flat plate-shaped brushite crystals simply consisted of preparing two solutions. Solution A was prepared as follows: 0.825 g of KH_2PO_4 (≥99.9%, Cat. No: 1.04873, Merck KGaA) was dissolved in 700 mL of distilled water, followed by the addition of 3.013 g of Na_2HPO_4 (≥99.9%, Cat. No: 1.06586, Merck KGaA), which resulted in a clear solution of pH 7.5 at RT. Solution B (of pH 6.4) was prepared by dissolving 4.014 g of $CaCl_2 \cdot 2H_2O$ in (≥99.9%, Cat. No: 1.02382, Merck KGaA) 200 mL of distilled water. Solution B was then rapidly added to solution A and the precipitates formed were aged for 80 min at RT, by continuous stirring at 500 rpm (final solution pH 5.3). Solids recovered by filtration from their mother liquors were dried overnight at 65°C to obtain 3.28 g of FP-shaped brushite powders.

Transformation of WL-shaped Brushite into OCP in DMEM Solution

Glass media bottles (100 mL-capacity) containing 50 mL of DMEM solutions (DMEM, High glucose 1X, Sterile, Product No: 21063-029, Gibco, Invitrogen, USA) were used. The composition of the DMEM solutions of this study were given elsewhere [22]. 1.0 g of WL-shaped brushite powder was placed in each bottle and the plastic caps of the bottles were sealed. The bottles were placed in a microprocessor-controlled static air oven whose temperature was adjusted to 36.5±0.1°C. The times for ageing the brushite powders in DMEM solutions were selected as 24, 48 h and one week. The DMEM solution of the one week sample was replenished with a fresh solution after 120 h. At the end of the specified ageing periods, the solids were recovered from the solutions by using a porcelain Buechner funnel and No. 2 Whatman filter paper, with vacuum filtration. The solids were washed with 500 mL of distilled water. Washed samples were left to dry overnight at 65°C to finally obtain DMEM-transformed powders with the following weights; 0.80±0.03 g, 0.81±0.03 g, 0.70±0.03 g for the 24 h-, 48 h- and one week-aged samples, respectively.

Transformation of FL-shaped Brushite into OCP in DMEM Solution

Glass media bottles (100 mL-capacity) containing 50 mL of DMEM solutions (DMEM, High glucose 1X, Sterile, Product No: 21063-029, Gibco, Invitrogen, USA) were used. 0.65 g of FP-shaped brushite powder was placed in each bottle and the plastic caps of the bottles were sealed. The bottles were placed in a microprocessor-controlled static air oven whose temperature was adjusted to 36.5±0.1°C. The times for ageing the FL-type brushite powders in DMEM solutions were selected as 48 h and one week. The DMEM solution of the one week sample was replenished with a fresh solution at every 48 hours. At the end of the specified ageing periods, the solids were recovered from the solutions by using a porcelain Buechner funnel and No. 2 Whatman filter paper, with vacuum filtration. The solids were washed with 500 mL of distilled water. Washed samples were left to dry overnight at 65°C to finally obtain DMEM-transformed powders with the following weights; 0.46±0.01 g and 0.45±0.01 g for the 48 h- and one week-aged samples, respectively.

Sample Characterization

All powder samples were characterized by using a powder X-ray diffractometer (Advance D8, Bruker AG, Karlsruhe, Germany) after being lightly grinded by an agate mortar and pestle. The diffractometer was operated with a Cu tube at 40 kV and 40 mA equipped with a monochromator. Samples were scanned with a step size of 0.02 and a preset time of 5 s. Scanning electron microscopy (EVO 40, Zeiss, Dresden, Germany) was used to evaluate the morphology of the powder samples. The samples were sputter-coated, prior to imaging, with a 25 nm-thick gold layer to impart electrical conductivity to the specimen surfaces. Fourier-transform infrared spectroscopy (Spectrum One, Perkin Elmer, USA) analyses were performed after mixing 1 mg of sample powders with 300 mg of KBr powder, followed by compacting those into a thin pellet in a stainless steel die of 1 cm inner diameter. FTIR data were recorded over the range of 4000 to 400 cm^{-1} with 128 scans.

RESULTS

The X-ray diffraction (XRD) and Fourier-transform infrared (FTIR) spectra of the two different brushite powders (denoted as FP and WL powders) synthesized in this study are given in Fig. 1a. The XRD spectra given in Fig. 1b depicted that the for the synthesis of WL-shaped brushite crystals, 90 s of mixing was not enough and the calcite powders remained still unreacted after 90 s, but in a time of stirring as short as 6 minutes brushite was forming in the aqueous $CaCO_3$-$NH_4H_2PO_4$ suspensions.

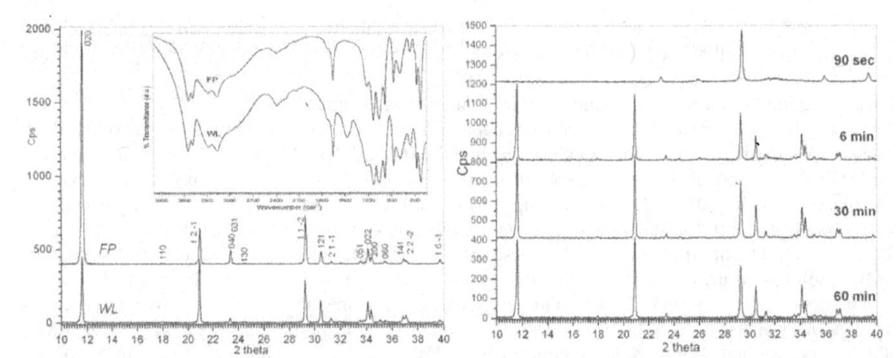

Fig. 1a XRD spectra of WL- (30 min stirring) and FP-type brushite powders; inset is depicting the FTIR spectrum of WL (30 min) and FP-type brushite

Fig. 1b XRD spectra of WL samples as a function of stirring time; 90 s trace was single-phase calcite, the other were pure brushite

Therefore, the selection of a mixing time between 6 and 60 minutes would be appropriate. There were no changes in the crystal size and shape of WL-type brushite powders if one increased the mixing time from 6 to 30 min or from 30 to 60 minutes. The scanning electron photomicrograph (SEM) of WL-type powders was shown in Fig. 1c.

Fig. 1c SEM photomicrograph WL-type brushite crystals

FP (flat-plate) powders consisted of large and flat plates of brushite, with a preferred growth along the (*020*) crystallographic planes and the relatively high X-ray intensities obtained from these planes were apparent in the top XRD trace of Fig. 1a. Those flat-plates were found to be thin (about 150 nm), had a width between 5 to 10 μm, and could elongate to about 70 to 80 μm. Flat-plate morphology is quite common to the precipitated brushite powders and could be frequently encountered in the literature on brushite. On the other hand, the WL (water lily, *Nymphaeaceae*) brushite powders of this study comprised dumbbells of water lily-shaped crystals. These crystals were again large, about 80-100 μm in size. For the WL-type brushite crystals (in comparison to the FP powders) almost equal X-ray intensities for the *(020)* and *(1 2 -1)* planes were registered as seen in Fig. 1a; i.e., the bottom XRD trace.

The FTIR inset shown in Fig. 1a disclosed that the WL powders contained a certain amount of unreacted $CaCO_3$, which was not detected by the XRD spectra of the same powders. The starting powder for the WL-type brushite was precipitated-chalk type $CaCO_3$ spindle-shaped particles. The SEM morphology of the $CaCO_3$ powders used in the current study was given elsewhere [23]. During the synthesis of WL-type brushite, these $CaCO_3$ (calcite) particles were chemically attacked by the acidic $H_2PO_4^-$ ions in solution, and a templated synthesis-type reaction followed that. The calcite particles were behaving as the template and the dumbbells of stacked brushite water lilies gradually formed in place of the original template. Since this was a room temperature aqueous reaction starting from the surface of the calcite template particles and advancing with time toward the cores of particles, it could be regarded as reasonable to have very small amounts (i.e., not detectable by the X-rays) of unreacted $CaCO_3$ at their cores. Such WL-shaped brushite crystals, to the best of our knowledge, were not reported prior to this study [22].

The XRD, FTIR and SEM data given in Fig. 2 denoted that the monetite ($CaHPO_4$) powders obtained from the 200 C-heating (for 2 h) of WL-brushite powders were single-phase and still preserved the WL morphology. Brushite-to-monetite conversion follows a simple dehydration reaction. The theoretical weight loss in the brushite-to-monetite conversion is 20.94%, which represents the loss of two water molecules. Our 200 C-heating runs always exhibited around 21% weight loss. Monetite powders with this WL-type particle morphology, to the best of our knowledge, were again not reported before. Commercially available monetite powders typically comprise rectangular prismatic or cubic particles (whose SEM morphology was reported elsewhere [24]).

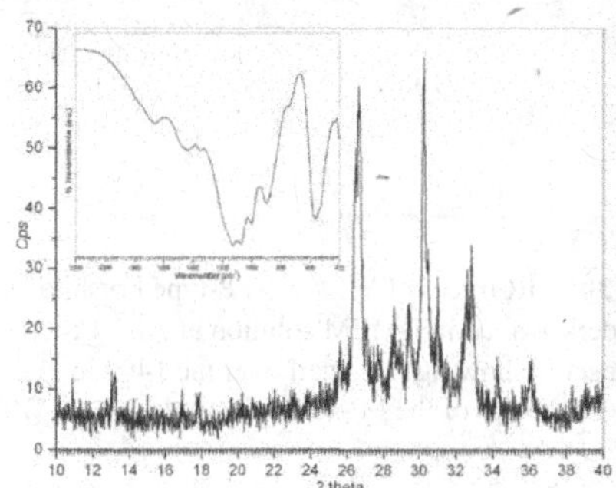

Fig. 2a XRD and FTIR (inset) spectra of monetite ($CaHPO_4$) powders of WL type

Fig. 2b Characteristic SEM photomicrograph of monetite ($CaHPO_4$) powders of WL morphology

Brushite crystallizes in the monoclinic space group *Cc* with the lattice parameters a=6.359, b=15.177, c=5.81 Å, β=118.54° [25]. Triclinic monetite has the following lattice parameters: a=6.910, b=6.627, c=6.998 Å, α=96.34°, β=103.82°, and γ=88.33° [25]. The experimentally-determined lattice parameters of brushite and monetite powders synthesized in this study differed very slightly (only in the third decimal place) from the above-mentioned literature values.

The analysis of FTIR data reproduced in the inset of Fig. 1a revealed the following IR frequencies (in cm^{-1}). The bands at 3544, 3491, 3290, and 3163 were due to the O-H stretching of water. The shoulder at 2955 was again that of O-H stretching. H_2O bending was recorded at 1653 cm^{-1}. The O-H in-plane bending was measured at 1219 cm^{-1}. PO stretching was observed at 1134, 1057, and 987 cm^{-1}. P-O(H) stretching was found at 876 cm^{-1} for the FP (flat-plates) sample, however, its shift to 871 cm^{-1} in the WL (water lily) sample together with the appearance of a carbonate band at 1472 cm^{-1} was indicative of unreacted calcite presence in the WL samples. H_2O libration was observed in both samples at 791 cm^{-1}. Finally, PO bending was recorded at 662, 576, 525 cm^{-1}. The observed IR band positions (of Figs. 1a and 2) were in close agreement with those reported by Xu *et al.* [26].

Both brushite powders were largely transformed into octacalcium phosphate (OCP, $Ca_8(HPO_4)_2(PO_4)_4\ 5H_2O$) upon one week of soaking in DMEM solution at 36.5 C, as shown by the X-ray diffraction data of Fig. 3a. WL samples still contained some unreacted brushite phase after 1 week in DMEM, but the FP samples were able to completely transform into OCP. Even after 48 h of ageing (pH dropping from the initial 7.4 to around 6.8) in the non-replenished DMEM solutions, OCP was the major phase in the FP samples. The lattice parameters of the triclinic OCP phase determined from the FP-1 week sample (Fig. 3a, top trace) were a=9.530, b=18.991, c=6.854 Å, α=92.30°, β=90.11°, and γ=79.94°, and they were in close agreement with those reported in ICDD PDF 026-1056 [27]. The FTIR data of the above samples were depicted in Fig. 3b, which were in good agreement with those previously reported by Wu and Nancollas [28], Suzuki *et al.* [29], LeGeros *et al.* [30], and LeGeros [31].

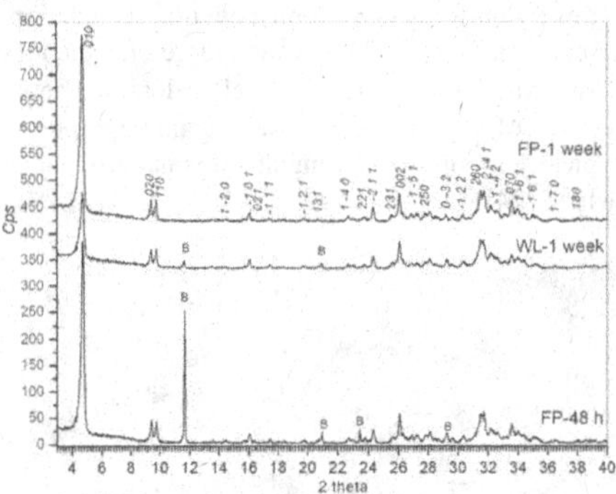

Fig. 3a XRD traces of WL- and FP-type brushite powders soaked in DMEM solution at 36.5 C (times indicated soaking periods); FP-1 week trace showing single-phase OCP together with the crystallographic indices of the OCP phase; letter B indicated the brushite peaks

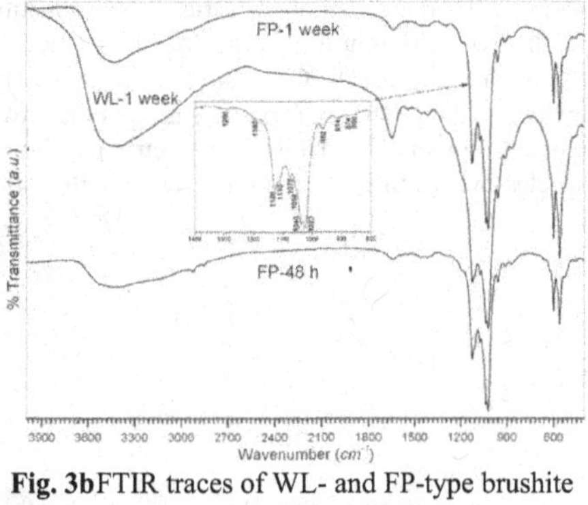

Fig. 3b FTIR traces of WL- and FP-type brushite powders soaked in DMEM solution at 36.5 C; the inset is showing the detail over the 1400 to 800 cm^{-1} range of the FP-1 week samples

The following IR bands were observed in the FTIR spectrum of FP-1 week sample shown in Fig. 3b. H-O-H or crystalline water of OCP was assigned to the wide band recorded over the range of 3700-3000 cm^{-1}. H_2O bending was at 1647 cm^{-1}. The P-OH bending modes originating from the HPO_4 groups of OCP were observed at 1296 and 1193 cm^{-1}. P-O in HPO_4 and PO_4 groups were recorded at 1126, 1110, 1075, 1058, 1040, 1023, 962, 628, 603, 560, 471, and 452 cm^{-1}. The P-OH stretching mode of HPO_4 groups was at 914 and 874 cm^{-1}. Finally, the HO-PO_3 bending mode in HPO_4 was found at 525 cm^{-1}. The IR band assignments of the FP-1 week samples of this study were in good agreement with those reported by LeGeros *et al.* [30, 31].

The changes occurred in the particle morphology of brushite powders aged in DMEM solutions at 36.5 C were followed by the SEM photomicrographs given in Figures 4a through 4d. It was found that one week was enough to fully convert the brushite crystals into octacalcium phosphate, as also supported by the FTIR and XRD data.

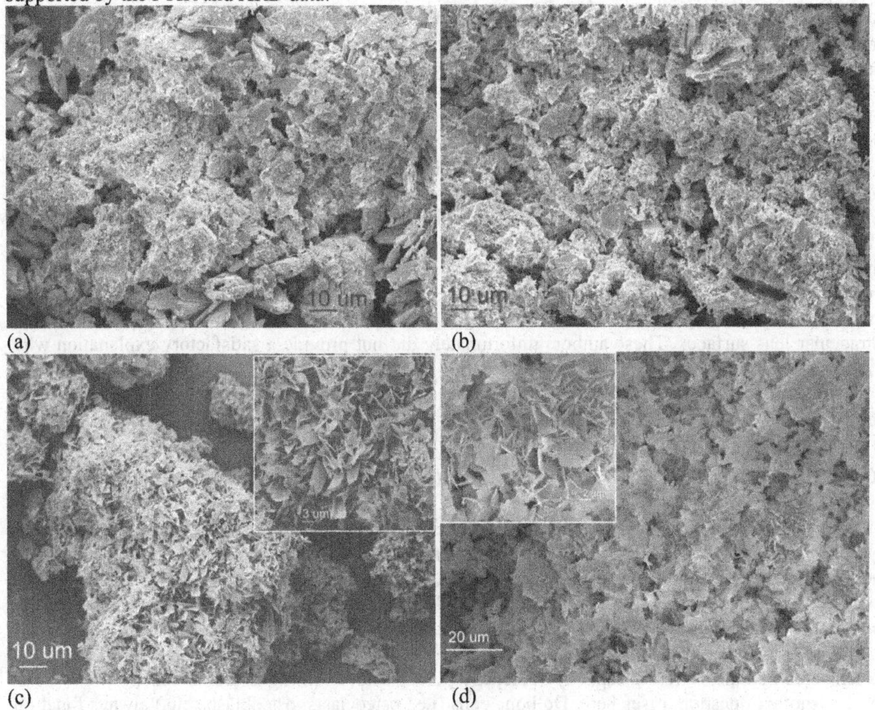

Fig. 4 SEM photomicrographs of samples soaked in DMEM solutions at 36.5 C; (a) WL – 24 h, (b) WL – 72 h, (c) WL – 1 week, (d) FP – 1 week

The DMEM solution used in this study was shown to be a convenient and robust medium to synthesize OCP powders in a static glass medium bottle, heated at 36.5 C, by starting with brushite powders synthesized in this study. It must be remembered that brushite powders soaked in SBF solutions at 37 C, under similar conditions, were only transforming into apatite [7-21].

In the current study, the DMEM solutions aged alone (i.e., without any brushite powders) in clean and sealed glass media bottles at 37 C for one week did not produce any precipitates within the bottles. This finding indicated that the DMEM solutions were not autogenously precipitating the OCP phase. In stark contrast to this, if one heated a volume of freshly prepared SBF solution, to be contained in a clean, sealed bottle (whether glass or plastic) at 36.5 C for one week, it would have produced lots of apatitic CaP precipitates by itself. How can one use such a medium to test bioactivity?

DISCUSSION

The major question resulting from this study is the following. If the brushite crystals soaked in Hepes-buffered DMEM solutions transform into OCP, whereas brushite crystals soaked in Tris-buffered SBF solutions transform into HA as reported in the previous studies [7-21], then which medium (DMEM or SBF) one needs to rely on as the correct bioactivity test for at least the brushite powders?

Octacalcium bis(hydrogenphosphate) tetraphosphate pentahydrate, typically referred to as octacalcium phosphate (OCP, $Ca_8(HPO_4)_2(PO_4)_4\ 5H_2O$), was proposed by Brown [32, 33] to be the precursor during the formation of apatitic calcium phosphate phase in teeth enamel mineralization and also in bone formation. OCP is a thermodynamically metastable phase with respect to hydroxyapatite (HA, $Ca_{10}(PO_4)_6(OH)_2$) [34], and the transformation from OCP to HA usually takes place rapidly, which is governed by the solution factors such as ion concentrations and pH. The log K_{SP} values of OCP and HA phases are -72.5 [35] and -117.1 [6], respectively. This significant difference between the solubilities of OCP and HA explains well why the mineralized portion of human hard tissues cannot be made of OCP alone. It seems like nature uses a transient phase like OCP as the initial step in reaching to the more durable HA crystallites in her biological biomineralization and maturation processes.

One must take into account that implanted intraocular lenses exhibited (*ex vivo*) calcification phenomena in certain cases, and very interestingly the calcified phase was OCP, but not HA [36]. Guan *et al.* [36] reported that both silicone and fatty acids, such as myristic, palmitic, stearic, arachidic, and behenic, had important roles in inducing OCP nucleation and growth on silicone-treated intraocular lens surfaces. These authors unfortunately did not provide a satisfactory explanation why HA was not observed *in lieu* of OCP.

Brushite transforms into OCP by following the below reactions;

$$10CaHPO_4\ 2H_2O \rightarrow Ca_8(HPO_4)_2(PO_4)_4\ 5H_2O + 2Ca^{2+} + 4H_2PO_4^- + 15H_2O \quad (2)$$

$$6CaHPO_4\ 2H_2O(s) + 2Ca^{2+} \rightarrow Ca_8(HPO_4)_2(PO_4)_4\ 5H_2O(s) + 7H_2O + 4H^+ \quad (3)$$

$$5CaHPO_4\ 2H_2O(s) + 3Ca^{2+} + HPO_4^{2-} \rightarrow Ca_8(HPO_4)_2(PO_4)_4\ 5H_2O(s) + 5H_2O + 4H^+ \quad (4).$$

OCP is a more basic calcium phosphate phase in comparison to brushite, and OCP shall further hydrolyze to convert itself into HA. The transformation of brushite into OCP should have been facilitated by heterogeneous nucleation and the role of possible crystal defects on the brushite surfaces could well be considered [37]. The generation of $H_2PO_4^-$ ions in reaction-2 would cause a readily measurable pH decrease [38]. This was why we observed a consistent pH decrease in our DMEM ageing solutions, from 7.4 to around 6.8 within the first 72-120 hours of ageing. It must be remembered that the buffering capacity of Hepes [39, 40] is not as strong as that of Tris [41].

Another question arises here. Do bone cells (i.e., osteoclasts, osteoblasts, etc.) always function and proliferate at a stable pH of 7.3 to 7.4 while they participate in the remodeling of bones? The answer to this question is negative and it is known that bone resorption (the first leg of healthy bone remodeling) is increased in metabolic acidosis [42], which can be mimicked experimentally by maintaining the cells at the extracellular pH of 6.5. Schilling *et al.* [43], for instance, found a more than four-fold increase in the number of osteoclasts compared to physiologic pH in a slightly acidic culture environment with an optimum between pH 6.9 and pH 7.1.

On the other hand, in uncorrected acidosis, the deposition of alkaline mineral (i.e., HA) in bone by osteoblasts (the second leg of healthy bone remodeling) is reduced, and osteoclast resorptive activity is increased in order to maximize the availability of OH^- ions in solution to counteract protons. Osteoblast alkaline phosphatase (ALP) activity peaks near the physiologic pH 7.4, but was found by Brandao-Burch *et al.* [44] to be reduced eight-fold at around pH 6.9. The same pH reduction is associated with two-and four-fold increases in Ca^{2+} and HPO_4^{2-} solubility for hydroxyapatite, respectively [44]. For more detailed analysis of the influence of medium pH on the activity of human osteoblasts in culture (such as alkaline phosphatase activity, lactate production, proline hydroxylation, DNA content and thymidine incorporation), interested readers may consult the article by Kaysinger and Ramp [45].

If, for example, a Tris-buffered SBF solution cannot allow its pH to drop from 7.4 to around 6.5-6.9, then it could not be possible for such a solution to mimic any osteoclast resorptive activity. Such a solution seems to be only programmed to heterogeneously nucleate nano-textured HA crystallites on the immersed substrates which do not cause a shift in the solution pH value towards 6.5 to 7. Hydroxylated surfaces ease this HA nucleation, and the pre-soaking of titanium coupons in heated solutions of NaOH or KOH prior to the SBF immersion constitutes a good example.

Calcification in cell culture media is a well-known phenomenon. For example, de Jonge *et al.* [45] very recently reported that soaking titanium and calcium phosphate-coated titanium substrates in α-MEM (supplemented with fetal calf or bovine serum (FBS), ascorbic acid, glycerophosphate and gentamycin) solutions at 37 C led to the formation of apatitic calcium phosphate calcification on those. However, de Jonge *et al.* [46] did not report the formation of OCP. The presence or absence of FBS in cell culture media exerts a significant difference on the calcification products.

There surely is a difference in the calcification potentials of α-MEM (minimum essentials medium Eagle, α-modification) and DMEM solutions, and this was best explored in the study of Coelho *et al.* [47] performed on the human osteoblastic cell cultures. DMEM is a less nutrient-rich medium with respect to amino acids and vitamins, although, nutrient concentrations are, on the whole, higher than those found in -MEM [47]. -MEM contains ascorbic acid, and less $NaHCO_3$ (27 mM) than that found in DMEM (44 mM). Coelho *et al.* [47] reported the formation of slightly higher amounts of apatitic calcium phosphate spherules (but not OCP) in DMEM solutions after culturing the osteoblasts for more than four weeks, by providing SEM photomicrographs and EDXS analysis results.

Interestingly, Price *et al.* [48] found in their calcification studies on rat aortas soaked in DMEM solutions alone were not able to form any calcification products; however, DMEM solutions supplemented with 1.5% human, bovine or rat blood serum were able to form apatitic calcification products easily on the same rat aortas. Price *et al.* [48] attributed this behaviour to the presence of a potent serum calcification factor (i.e., the noncollagenous serum protein fetuin) in serum.

Therefore, in comparing the Coelho *et al.* [47] and Price *et al.* [48] studies one should realize that in the first study the DMEM solutions contained osteoblasts and the alkaline phosphatase (ALP) released from the osteoblasts would appear in the medium, whereas in the latter study there were no osteoblasts to secrete ALP and hence the serum calcification factors.

Until now, OCP crystallization was more or less considered to take place in aqueous solutions containing monocarboxylates, such as acetate (or formate) ions; this was probably due to the quite influential papers by Newesely [49] and LeGeros [31] on OCP synthesis. These synthesis procedures involved the mixing of calcium nitrate solutions with sodium acetate solutions or calcium acetate solutions with sodium phosphate solutions, respectively. LeGeros procedure envisaged the precipitation process to be performed between 60 and 80 C [31]. Liu *et al.* [50] slightly modified the Newesely procedure of OCP synthesis, mixed calcium nitrate and disodium hydrogen phosphate solutions in a sodium acetate solution, continued the precipitation-maturation process at 45 C for 48 h, but were not able to obtain single-phase OCP powders according to their FTIR spectra. Brown *et al.* [32] have historically been the first to hydrolyze brushite powders in a concentrated sodium acetate solution, and the same procedure was recently repeated by Monma *et al.* [51]. Our study, on the other hand, demonstrated a new route which totally eliminated the use of quite concentrated (0.2 M [50] or 0.5 M [51]) acetate solutions for synthesizing the OCP powders.

The absence of any previous studies related to the ageing/immersion of brushite crystals or powders, at the human body temperature of 36.5 C, in a DMEM solution containing amino acids, the present authors are somewhat compelled to search for studies performed on titanium by using other calcification solutions.

Wen *et al.* [53] observed the formation of an OCP-like phase on the surface of HCl-H_2SO_4 and NaOH-treated commercially pure titanium immersed into a Tris/HCl-buffered and Mg- and HCO_3-free supersaturated calcification solution, SCS, having a Ca/P molar ratio of 1.67. This solution was, therefore, different from the popular SBF solutions also in terms of its Ca/P molar ratio and its degree of supersaturation with respect to apatite nucleation. The absence of Mg^{2+} and HCO_3^- ions in the solution described by Wen *et al.* [52] and the observation of OCP in place of single-phase apatite must be underlined. Wen and Moradian-Oldak [53] later reported the formation of OCP (instead of apatite) on titanium surfaces immersed in Mg- and HCO_3-free and Tris-buffered supersaturated calcification solutions (SCS) both without and with bovine serum albumin or murine amelogenin. These two reports [52, 53], contributed by the same first author, seemed to assert a significant difference in terms of the phase nature of the calcium phosphate deposited by the SCS (Mg- and HCO_3-free) and SBF (Mg- and HCO_3-containing) solutions. The same Mg- and HCO_3-free SCS solution, originally reported in the Wen *et al.* [52] paper, was then reproduced and reported in a number of seemingly follow-up studies to form OCP deposits on titanium [54, 55] or HA or HA-TCP bioceramics [56].

However, the Hepes-buffered DMEM solution used in our study contained both Mg^{2+} and HCO_3^- ions, and was still able to completely transform the immersed brushite powders into OCP in 7 days. Therefore, the crystallization of OCP from a Tris-buffered SCS (without Mg and HCO_3) or Hepes-buffered DMEM solution cannot be simply attributed to or explained by the absence or presence of Mg^{2+} and HCO_3^- ions. The main question here should again be focused on the pH-stability of solutions buffered by using Tris or Hepes. Wen *et al.* [52] reported that the pH of their SCS solutions started from 7.4 and gradually dropped to around 7.2 within the first 16 h of immersion. This rapid drop in pH should be the reason for forming OCP instead of HA on their titanium coupons.

A possible answer to this issue was provided by an article of Serro and Saramago [57]. Serro and Saramago [57] prepared a new solution designated as SBF0, with the same composition of Kokubo SBF [1] but without the buffer Tris. The buffer TRIS, present in SBF, is known to form soluble complexes with several cations, including the most important Ca^{2+}. This would in turn help to reduce the concentration of free Ca^{2+} ions in SBF with respect to an SBF-like solution without Tris [57]. The pH of DMEM solutions (containing the brushite powders) of the current study decreased from 7.4 to around 6.77 by the end of the first 72 hours at 36.5 C, then gradually rose to 6.83 over the following 48 hours of immersion. The pH values of the brushite-containing DMEM solutions were found to rise to 7.0 at the end of 7 days at 36.5 C. This meant that the Hepes-buffered DMEM solutions were not able to maintain the initial pH of 7.4 when they had the slightly acidic brushite powders in them. The SBF-without-Tris solutions of Serro and Saramago [57], on the other hand, aged at 37 C were found to exhibit pH values increasing from 7.4 to values of 8.5 (while precipitating HA) in the course of 7 days, whereas the same authors reported the corresponding pH variations for Tris-buffered SBF solutions to be below 0.1 within the same timeframe. Does this mean that maintaining the pH value at or not below 7.4 would be essential for producing HA from such solutions?

The morphology of the OCP crystals deposited on titanium, HA or HA-TCP samples [52-56] of previous studies were quite similar to those shown in Figs. 4c and 4d.

The influence of amino acids present in MEM solutions, in direct comparison to HBSS (Hanks' balanced salt solution) [58], was studied by Hiromoto *et al.* [59] in terms of the amount of calcium phosphate deposited on titanium coupons immersed in HBSS (Hanks' balanced salt solution) or MEM solutions over 7 days. HBSS solutions do not contain amino acids and any Tris or Hepes buffers [58]. The presence of biomolecules (in the case of MEM solutions) was found to decrease the amount of calcium phosphate deposited [59]. One of the most important contributions of the Hiromoto *et al.* [59] study had been the quantitative measurement of the Ca, P, C, N, and O amounts of the surface oxide films of titanium specimens immersed in HBSS and MEM solutions by using X-ray photoelectron spectroscopy. Quite small amounts of carbon and nitrogen (and especially nitrogen) detected in the calcium phosphate layers coated on titanium were originating directly from the amino acids present in the MEM solutions. Hiromoto *et al.* [59] study thus provided a strong evidence for the adsorption and incorporation of biomolecules, supplied by the amino acids of MEM solutions, in the newly forming calcium phosphate phases in such biocompatible media.

Although the experimental scope of our study was not wide enough to include the investigation of the influence of biomolecules (and their adsorption) on the brushite-to-OCP transformation occurring in DMEM at 36.5 C, it shall be not so unsafe to assume the adsorption of biomolecules on our OCP crystals.

The role of bovine serum albumin (BSA) on the crystallization of OCP on type I collagen of bovine origin in metastable supersaturated solutions of pH=6.5 at 37 C was studied by Combes *et al.* [60]. BSA was found to strongly influence the shape of the OCP crystals at the quite high level of 40 g/L, resulting in smaller crystals with curved edges [60]. The amino acids present in our DMEM solutions did not exhibit such an effect on the crystal morphology of OCP as seen in Figs. 4c and 4d. The SEM photomicrographs shown in Figs. 4c and 4d resembled to those supplied by Combes *et al.* [60] when they synthesized their samples in the absence of BSA.

To summarize, if the pH of a physiologic medium (either Hepes-buffered DMEM or Tris-buffered SCS [51]) is dropping to around 7.2-6.8 during hydrothermal ageing, then such a solution would be prone to nucleate crystals of OCP, rather than those of HA.

DMEM solution was found to have the ability of transforming brushite powders into OCP at 36.5 C within 1 week; OCP is a biological calcium phosphate phase; OCP is indeed the precursor to HA in enamel and bone formation; therefore, the DMEM solutions can be used to test the so-called bioactivity of brushite powders as confidently as SBF solutions. DMEM solutions, which are commonly used in cell culture [61-63], can be regarded as a feasible alternative to using SBF solutions in the so-called *in vitro* bioactivity testing of synthetic biomaterials.

In regard to follow-up studies, the transformation of brushite powders to OCP at the human body temperature, if performed in a cell culture solution such as DMEM, could also prove to be a viable option to synthesize OCP-based biomaterials in solutions, which can be specifically loaded with bioactive molecules, certain proteins and growth factors.

The current study has also offered a new synthesis route to prepare $CaHPO_4 \cdot 2H_2O$ powders with an unprecedented morphology, i.e., water lily(WL)-like brushite dumbbells. The main advantage of the WL (water lily)-type brushite or WL-type monetite powders will be the following. These powders were produced in simple aqueous systems which did not contain any organic or polymeric substances or surfactants. For example, the very recent use of cetyltrimethylammonium bromide (CTAB, a cationic hydrophobic detergent), by Ruan *et al.* [64], in preparing monetite powders with a flower-like morphology (i.e., a morphology outside the common flat plate-type observed in brushite synthesis) could only be regarded as an interesting attempt, but the resultant monetite powders must surely be suspected of containing the residues of CTAB used during synthesis. CTAB, for instance, is already known to cause chronic toxicity upon its digestion [65, 66]. The total absence of organics in the production does eliminate any toxicity concerns about the possible organic residues (even at the ppm levels) to be present in such calcium phosphate-based bone substitute biomaterials.

A new family of calcium phosphate cements with the final setting product being brushite (in stark contrast to more common hydroxyapatite cements) was developed within the last decade and they were claimed to show increased *in vivo* resorbability [67-69]. The investigation of the hydrothermal transformation behavior of brushite at 36.5 C in a cell culture medium, such as DMEM, was expected to increase the level of scientific understanding on the bioactivity of brushite-based biomaterials.

CONCLUSIONS

(1) A new chemical process was suggested for the robust and economical synthesis of brushite ($CaHPO_4 2H_2O$) powders with a unique water lily-like morphology. The new process involved the simple stirring of an aqueous suspension of precipitated calcite ($CaCO_3$) powders and dissolved ammonium dihydrogen phosphate ($NH_4H_2PO_4$), the suspension being free of any organic additives, at room temperature (24±1°C), from 6 to 60 minutes.

(2) The results showed that it was possible to preserve the water lily-like particle morphology of brushite even though the powders were later converted to monetite ($CaHPO_4$) by heating at 200 C. Monetite powders with water lily-shaped crystals were produced for the first time.

(3) This study offered a simple procedure for producing octacalcium phosphate ($Ca_8(HPO_4)_2(PO_4)_4 5H_2O$) powders, upon straightforward immersion of brushite powders in DMEM solutions (pH 7.4) at the human body temperature of 36.5°C. The highest temperature of processing hereby used in the manufacture of bulk, well-crystallized OCP powders was 36.5°C.

Notes: *Certain commercial equipment, instruments, solutions or chemicals are only identified in this paper to foster understanding. Such identification does not imply recommendation or endorsement by the authors, nor does it imply that the equipment or materials identified are necessarily the best available for the purpose.*

REFERENCES

[1] T. Kokubo, Surface Chemistry of Bioactive Glasses, *J. Non-Cryst. Solids*, **120**, 138-151 (1990).
[2] D. Bayraktar and A.C. Tas, Chemical Preparation of Carbonated Calcium Hydroxyapatite Powders at 37 C in Urea-containing Synthetic Body Fluids, *J. Eur. Ceram. Soc.*, **19**, 2573-2579 (1999).
[3] H.M. Kim, K. Kishimoto, F. Miyaji, T. Kokubo, T. Yao, Y. Suetsugu, J. Tanaka, and T. Nakamura, Composition and Structure of Apatite Formed on Organic Polymer in Simulated Body Fluid with a High Content of Carbonate Ion, *J. Mater. Sci. Mater. M.*, **11**, 421-426 (2000).
[4] M. Bohner and J. Lemaitre, Can Bioactivity be Tested in vitro with SBF Solution? *Biomaterials*, **30**, 2175-2179 (2009).
[5] T. Kokubo and H. Takadama, How Useful Is SBF in Predicting in vivo Bone Bioactivity? *Biomaterials*, **27**, 2907-2915 (2006).
[6] F.C.M. Driessens and R.M.H. Verbeeck, *Biominerals*, pp. 37-59, CRC Press, Boca Raton, FL, 1990.
[7] M. Kumar, H. Dasarathy, and C. Riley, Electrodeposition of Brushite Coatings and Their Transformation to Hydroxyapatite in Aqueous Solutions, *J. Biomed. Mater. Res.*, **45**, 302-310 (1999).
[8] D. Walsh and J. Tanaka, Preparation of Bone-like Apatite Foam Cement, *J. Mater. Sci. Mater. M.*, **12**, 339-343 (2001).
[9] L. Grondahl, F. Cardona, K. Chiem, E. Wentrup-Byrne, and T. Bostrom, Calcium Phosphate Nucleation on Surface-modified PTFE Membranes, *J. Mater. Sci. Mater. M.*, **14**, 503-510 (2003).
[10] S.J. Lin, R.Z. LeGeros, and J.P. LeGeros, Adherent Octacalcium Phosphate Coating on Titanium Alloy using Modulated Electrochemical Deposition Method, *J. Biomed. Mater. Res.*, **66A**, 819-828 (2003).
[11] C.Y. Kim and H.B. Lim, Hardening and Hydroxyapatite Formation on Bioactive Glass and Glass-ceramic Cement, *Key Eng. Mater.*, **254-2**, 305-308 (2004).
[12] H.S. Azevedo, I.B. Leonor, C.M. Alves, and R.L. Reis, Incorporation of Proteins and Enzymes at Different Stages of the Preparation of Calcium Phosphate Coatings on a Degradable Substrate by a Biomimetic Methodology, *Mater. Sci. Eng. C*, **25**, 169-179 (2005).
[13] X. Lu and Y. Leng, Theoretical Analysis of Calcium Phosphate Precipitation in Simulated Body Fluid, *Biomaterials*, **26**, 1097-1108 (2005).

[14] D.J.T. Hill-Zainuddin, T.V. Chirila, A.K. Whittaker, and A. Kemp, Experimental Calcification of HEMA-based Hydrogels in the Presence of Albumin and a Comparison to the in vivo Calcification, *Biomacromolecules*, **7**, 1758-1765 (2006).
[15] J. Pena, I. Barba-Izquierdo, A. Martinez, and M. Vallet-Regi, New Method to Obtain Chitosan/Apatite Materials at Room Temperature, *Solid State Sciences*, **8**, 513-519 (2006).
[16] T. Anada, T. Kumagai, Y, Honda, T. Masuda, R. Kamijo, S. Kamakura, and O. Suzuki, Dose-dependent Osteogenic Effect of Octacalcium Phosphate on Mouse Bone Marrow Stromal Cells, *Tissue Eng.*, **14**, 965-978 (2008).
[17] C. Knabe, A. Houshmand, G. Berger, P. Ducheyne, R. Gildenhaar, I. Kranz, and M. Stiller, Effect of Rapidly Resorbable Bone Substitute Materials on the Temporal Expression of the Osteoblastic Phenotype in vitro, *J. Biomed. Mater. Res.*, **84A**, 856-868 (2008).
[18] F. Yang, J.G.C. Wolke, and J.A. Jansen, Biomimetic Calcium Phosphate Coating on Electrospun Poly(epsilon-caprolactone) Scaffolds for Bone Tissue Engineering, *Chem. Eng. J.*, **137**, 154-161 (2008).
[19] J.A. Juhasz, S.M. Best, A.D. Auffret, and W. Bonfield, Biological Control of Apatite Growth in Simulated Body Fluid and Human Blood Serum, *J. Mater. Sci. Mater. M.*, **19**, 1823-1829 (2008).
[20] L.P. Xu, E.L. Zhang, and K. Yang, Phosphating Treatment and Corrosion Properties of Mg-Mn-Zn Alloy for Biomedical Application, *J. Mater. Sci. Mater. M.*, **20**, 859-867 (2009).
[21] A. Rakngarm and Y. Mutoh, Electrodeposition of Calcium Phosphate Film on Commercial Pure Titanium and Ti-6Al-4V in Two Types of Electrolyte at Room Temperature, *Mater. Sci. Eng. C*, **29**, 275-283 (2009).
[22] S. Mandel and A.C. Tas, Brushite ($CaHPO_4$ $2H_2O$) to Octacalcium Phosphate ($Ca_8(HPO_4)_2(PO_4)_4$ $5H_2O$) Transformation in DMEM Solutions at 36.5 C, *Mater. Sci. Eng. C*, **30**, 245-254 (2010).
[23] A.C. Tas, Porous, Biphasic $CaCO_3$-Calcium Phosphate Biomedical Cement Scaffolds from Calcite ($CaCO_3$) Powder, *Int. J. Appl. Ceram. Technol.*, **4**, 152-163 (2007).
[24] S. Jalota, S.B. Bhaduri, and A.C. Tas, Using a Synthetic Body Fluid (SBF) Solution of 27 mM HCO_3^- to Make Bone Substitutes More Osteointegrative, *Mater. Sci. Eng. C*, **28**, 129-140 (2008).
[25] L. Tortet, J.R. Gavarri, G. Nihoul, and A.J. Dianoux, Study of Protonic Mobility in $CaHPO_4$ $2H_2O$ (brushite) and $CaHPO_4$ (monetite) by Infrared Spectroscopy and Neutron Scattering, *J. Solid State Chem.*, **132**, 6-16 (1997).
[26] J. Xu, I.S. Butler, and D.F.R. Gilson, FT-Raman and High-Pressure Infrared Spectroscopic Studies of Dicalcium Phosphate Dihydrate ($CaHPO_4$ $2H_2O$) and Anhydrous Calcium Phosphate ($CaHPO_4$), *Spectrochimica Acta A*, **55**, 2801-2809 (1999).
[27] ICDD PDF: International Centre for Diffraction Data, Powder Diffraction File. PA, USA.
[28] W. Wu and G.H. Nancollas, Nucleation and Crystal Growth of Octacalcium Phosphate on Titanium Oxide Surfaces, *Langmuir*, **13**, 861-865 (1997).
[29] O. Suzuki, S. Kamakura, T. Katagiri, M. Nakamura, B. Zhao, Y. Honda, and R. Kamijo, Bone Formation Enhanced by Implanted Octacalcium Phosphate involving Conversion into Ca-deficient Hydroxyapatite, *Biomaterials*, **27**, 2671-2681 (2006).
[30] R.Z. LeGeros, G. Daculsi, I. Orly, T. Abergas, and W. Torres, Solution-mediated Transformation of Octacalcium Phosphate (OCP) to Apatite, *Scanning Microscopy*, **3**, 129-138 (1989).
[31] R.Z. LeGeros, Preparation of Octacalcium Phosphate (OCP): A Direct Fast Method, *Calcified Tissue Int.*, **37**, 194-197 (1985).
[32] W.E. Brown, J.P. Smith, J.R. Lehr, and A.W. Frazier, Crystallographic and Chemical Relations between Octacalcium Phosphate and Hydroxyapatite, *Nature*, **196**, 1048-1050 (1962).
[33] E.C. Moreno, T.M. Gregory and W.E. Brown, Solubility of $CaHPO_4$ $2H_2O$ and Formation of Ion Pairs in System $Ca(OH)_2$-H_3PO_4-H_2O at 37.5 C, *J. Res. Natl. Bur. Std.*, **70A**, 545-552 (1966).

[34] M. Iijima, Formation of Octacalcium Phosphate in vitro, *Monogr. Oral Sci.*, **18**, 17-49 (2001).
[35] L.J. Shyu, L. Perez, S.L. Zawacki, J.C. Heughebaert, and G.H. Nancollas, The Solubility of Octacalcium Phosphate at 37 C in the System $Ca(OH)_2$-H_3PO_4-KNO_3-H_2O, *J. Dent. Res.*, **62**, 398-400 (1983).
[36] X. Guan, R. Tang and G.H. Nancollas, The Potential Calcification of Octacalcium Phosphate on Intraocular Lens Surfaces, *J. Biomed. Mater. Res.*, **71A**, 488-496 (2004).
[37] H.E.L. Madsen, The Growth of Dicalcium Phosphate Dihydrate on Octacalcium Phosphate at 25 C, *J. Cryst. Growth*, **80**, 450-452 (1987).
[38] H.E.L. Madsen, Influence of Foreign Metal Ions on Crystal Growth and Morphology of Brushite ($CaHPO_4$ $2H_2O$) and its Transformation to Octacalcium Phosphate and Apatite, *J. Cryst. Growth*, **310**, 2602-2612 (2008).
[39] R.Z. Sabirov, J. Prenen, G. Droogmans, and B. Nilius, Extra- and Intracellular Proton-binding Sites of Volume-regulated Anion Channels, *J. Membrane Biol.*, **177**, 13-22 (2000).
[40] L. Fulop, G. Szigeti, J. Magyar, N. Szentandrassy, T. Ivanics, Z. Miklos, L. Ligeti, A. Kovacs, G. Szenasi, L. Csernoch, P.P. Nanasi, and T. Banyasz, Differences in Electrophysiological and Contractile Properties of Mammalian Cardiac Tissues Bathed in Bicarbonate- and Hepes-buffered Solutions, *Acta Physiol. Scand.*, **178**, 11-18 (2003).
[41] J. Pratt, J.D. Cooley, C.W. Purdy, and D.C. Straus, Lipase Activity from Strains of Pasteurella Multocida, *Curr. Microbiol.*, **40**, 306-309 (2000).
[42] T. Nordstrom, L.D. Shrode, O.D. Rotstein, R. Romanek, T. Goto, J.N.M. Heersche, M.F. Manolson, G.F. Brisseau, and S. Grinstein, Chronic Extracellular Acidosis Induces Plasmalemmal Vacuolar type H^+ ATPase Activity in Osteoclasts, *J. Biol. Chem.*, **272**, 6354-6360 (1997).
[43] A.F. Schilling, W. Linhart, S. Filke, M. Gebauer, T. Schinke, J.M. Rueger, and M. Amling, Resorbability of Bone Substitute Biomaterials by Human Osteoclasts, *Biomaterials*, **25**, 3963-3972 (2004).
[44] A. Brandao-Burch, J.C. Utting, I.R. Orriss, and T.R. Arnett, Acidosis Inhibits Bone Formation by Osteoblasts in vitro by Preventing Mineralization, *Calcified Tissue Int.*, **77**, 167-174 (2005) .
[45] K.K. Kaysinger and W.K. Ramp, Extracellular pH Modulates the Activity of Cultured Human Osteoblasts, *J. Cell. Biochem.*, **68**, 83-89 (1998).
[46] L.T. de Jonge, J.J.J.P. van den Beucken, S.C.G. Leeuwenburgh, A.A.J. Hamers, J.G.C. Wolke, and J.A. Jansen, The Osteogenic Effect of Electrosprayed Nanoscale Collagen/Calcium Phosphate Coatings on Titanium, *Biomaterials*, **31**, 2461-2469 (2010).
[47] M.J. Coelho, A.T. Cabral, and M.H. Fernandes, Human Bone Cell Cultures in Biocompatiblity Testing. Part I: Osteoblastic Differentiation of Serially Passaged Human Bone Marrow Cells Cultured in -MEM and in DMEM, *Biomaterials*, **21**, 1087-1094 (2000).
[48] P.A. Price, W.S. Chan, D.M. Jolson, and M.K. Williamson, The Elastic Lamellae of Devitalized Arteries Calcify when Incubated in Serum – Evidence for a Serum Calcification Factor, *Arterioscl. Throm. Vas.*, **26**, 1079-1085 (2006).
[49] E. Hayek and H. Newesely, Pentacalcium Monohydroxyorthophosphate, *Inorg. Syn.*, **7**, 63-65 (1963).
[50] Y. Liu, P.R. Cooper, J.E. Barralet, and R.M. Shelton, Influence of Calcium Phosphate Crystal Assemblies on the Proliferation and Osteogenic Gene Expression of Rat Bone Marrow Stromal Cells, *Biomaterials*, **28**, 1393-1403 (2007).
[51] H. Monma and T. Kamiya, Preparation of Hydroxyapatite by the Hydrolysis of Brushite, *J. Mater. Sci.*, **22**, 4247-4250 (1987).
[52] H.B. Wen, J.G.C. Wolke, J.R. de Wijn, Q. Liu, F.Z. Cui, and K. de Groot, Fast Precipitation of Calcium Phosphate Layers on Titanium Induced by Simple Chemical Treatments, *Biomaterials*, **18**, 1471-1478 (1997).

[53] H.B. Wen and J. Moradian-Oldak, Modification of Calcium Phosphate Coatings on Titanium by Recombinant Amelogenin, *J. Biomed. Mater. Res.*, **64A**, 483-490 (2003).
[54] F. Barrere, P. Layrolle, C.A. van Blitterswijk, and K. de Groot, Biomimetic Coatings on Titanium: A Crystal Growth Study of Octacalcium Phosphate, *J. Mater. Sci. Mater. M.*, **12**, 529-534 (2001).
[55] F. Barrere, C.M. van der Valk, R.A.J. Dalmeijer, G. Meijer, C.A. van Blitterswijk, K. de Groot, and P. Layrolle, Osteogenecity of Octacalcium Phosphate Coatings Applied on Porous Metal Implants, *J. Biomed. Mater. Res.*, **66A**, 779-788 (2003).
[56] P. Habibovic, C.M. van der Valk, C.A. van Blitterswijk, G. Meijer, and K. de Groot, Influence of Octacalcium Phosphate Coating on Osteoinductive Properties of Biomaterials, *J. Mater. Sci. Mater. M.*, **15**, 373-380 (2004).
[57] A.P. Serro and B. Saramago, Influence of Sterilization on the Mineralization of Titanium Implants Induced by Incubation in Various Biological Model Fluids, *Biomaterials*, **24**, 4749-4760 (2003).
[58] J.H. Hanks and R.E. Wallace, Relation of Oxygen and Temperature in the Preservation of Tissues by Refrigeration, *Proc. Soc. Exp. Biol. Med.*, **71**, 196-199 (1949).
[59] S. Hiromoto, T. Hanawa, and K. Asami, Composition of Surface Oxide Film of Titanium with Culturing Murine Fibroblasts L929, *Biomaterials*, **25**, 979-986 (2004).
[60] C. Combes, C. Rey, and M. Freche, In vitro Crystallization of Octacalcium Phosphate on Type I Collagen: Influence of Serum Albumin, *J. Mater. Sci. Mater. M.*, **10**, 153-160 (1999).
[61] J.L. Ong, D.R. Villarreal, R. Cavin, and K. Ma, Osteoblast Responses to As-deposited and Heat-treated Sputtered CaP Surfaces, *J. Mater. Sci. Mater. M.*, **12**, 491-495 (2001).
[62] J.E. Gough, I. Notingher, and L.L. Hench, Osteoblast Attachment and Mineralized Nodule Formation on Rough and Smooth 45S5 Bioactive Glass Monoliths, *J. Biomed. Mater. Res.*, **68A**, 640-650 (2004).
[63] T.J. Webster and J.U. Ejiofor, Increased Osteoblast Adhesion on Nanophase Metals: Ti, TI6Al4V, and CoCrMo, *Biomaterials*, **25**, 4731-4739 (2004).
[64] Q.C. Ruan, Y.C. Zhu, Y. Zeng, H.F. Qian, J.W. Xiao, F.F. Xu, L.L. Zhang, and D.H. Zhao, Ultrasonic-irradiation-assisted Oriented Assembly of Ordered Monetite Nanosheets Stacking, *J. Phys. Chem. B*, **113**, 1100-1106 (2009).
[65] B. Isomaa, J. Reuter, and B.M. Djupsund, Subacute and Chronic Toxicity Cetyltrimethylammonium Bromide (CTAB), A Cationic Surfactant, in Rat, *Arch. Toxicol.*, **35**, 91-96 (1976).
[66] M.J. Heffernan, S.P. Kasturi, S.C. Yang, B. Pulendran, and N. Murthy, The Stimulation of CD8(+) T Cells by Dendritic Cells Pulsed with Polyketal Microparticles Containing Ion-paired Protein Antigen and Poly(inosinic acid)-Poly(cytidylic acid), *Biomaterials*, **30**, 910-918 (2009).
[67] B. Flautre, C. Maynou, J. Lemaitre, P. Van Landuyt, and P. Hardouin, Bone Colonization of -TCP Granules Incorporated in Brushite Cements, *J. Biomed. Mater. Res.*, **63B**, 413-417 (2002).
[68] M. Bohner, F. Theiss, D. Apelt, W. Hirsiger, R. Houriet, G. Rizzoli, E. Gnos, C. Frei, J.A. Auer, and B. von Rechenberg, Compositional Changes of a Dicalcium Phosphate Dihydrate Cement after Implantation in Sheep, *Biomaterials*, **24**, 3463-3474 (2003).
[69] M. Bohner, U. Gbureck, and J.E. Barralet, Technological Issues for the Development of More Efficient Calcium Phosphate Bone Cements: A Critical Assessment, *Biomaterials*, **26**, 6423-6429 (2005).

HYDROXYAPATITE SCAFFOLDS FOR BONE TISSUE ENGINEERING WITH CONTROLLED POROSITY AND MECHANICAL STRENGTH

Vincenzo M. Sglavo[1], Marzio Piccinini[1], Andrea Madinelli[1], Francesco Bucciotti[2]

[1]University of Trento, Department of Materials Engineering and Industrial Technologies, Trento, Italy
[2]Eurocoating SpA, Cirè-Pergine Valsugana (TN), Italy

ABSTRACT

Hydroxyapatite (HA) macro-porous scaffolds suitable for bone tissue engineering applications with controlled porosity and mechanical strength were realized and characterized in the present work. The foam replica method was employed for the production of the ceramic components and processing conditions were varied in order to obtain materials with different porosity, macro-pores size and compressive strength. Porosity from 70 to 80% and strength from 0.5 to 2 MPa were obtained. An interconnected porous structure with pores size between 100 and 500 μm was obtained, this allowing cells penetration, tissue in-growth and vascularization as required in bone tissue engineering.

INTRODUCTION

Calcium phosphate and hydroxyapatite (HA) ceramics are considered among the most promising materials for bone replacement because of their bone-like chemical composition and mechanical properties. For bone tissue engineering it is important to improve different parameters such as porosity, pore size, shape and interconnectivity [1]. Porosity is necessary *in vivo* for the bone tissue ingrowth since it allows migration and proliferation of osteoblasts and mesenchymal cells and matrix deposition in the empty spaces. The interconnectivity plays a fundamental role for blood vessels invasion because a bone vascularization coordinates the activity of bone cells and their migration for bone remodeling [2]. In addition, mechanical properties of the scaffold play and additional fundamental role especially in avoiding crack formation during the use of the component.
In the present work HA scaffolds with different porosity and mechanical strength were produced by varying the conditions of the sponge-replica production process.

EXPERIMENTAL PROCEDURES

HA scaffolds were fabricated by the foam-replica method [3-4]. Two suspensions, labeled as A and B, with different viscosity were produced by mixing commercial HA micro-powder, deionized water, dispersant and binder with an automatic mixer to ensure complete homogenization. The steady-state viscosity was equal to about 5 Pa s and 0.05 Pa s for suspension A and B, respectively. Cylindrical (diameter ≈ 10 mm, height ≈ 20 mm) commercial polyurethane sponge with porosity of 97% and pore dimensions in the range 500-1200 μm were impregnated several times with the high viscosity suspension and then with the other one, thus depositing different amount of solid on the foam (low, medium, high). Subsequently, the material in excess was removed and the sponge was dried and subjected to thermal treatment for the removal of the organics, the burn-out of the polymeric sponge and successive HA sintering.

The morphology and porosity of the HA sponges were analyzed by optical microscope (Stemi 2000 C, Zeiss) and SEM (JSM-5500, Jeol), and the average pore size was calculated by Image J software. The compressive strength tests were performed using a servo-hydraulic machine (MTS System 810), according to ASTM D1621 norm. XRD patterns were recorded on a Rigaku Dmax III diffractometer using a Cu kα radiation, operating at 40 kV and 30 mA.

RESULTS AND DISCUSSION

The microstructure of the obtained hydroxyapatite scaffolds with different solid loadings is shown in Figure 1. The structure is characterized by a highly interconnected macro-porous architecture.

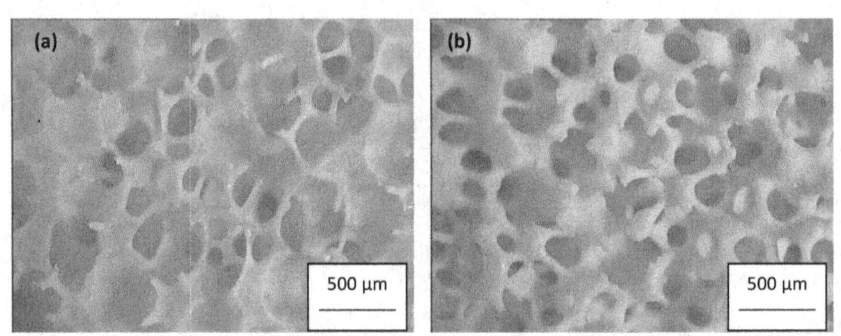

Figure 1. Optical micrograph of porous HA scaffolds produced with low (a) and high (b) solid loading.

SEM observations (Figures 2 and 3) allowed also the measurement of macro- and micro-pore sizes that range between 100 and 500 μm and 1 to 10 μm in low and high solid loading, respectively. In the low solid loading scaffolds pore struts appear sharper, perfectly following the PU sponge morphology, which presents some acicular asperities that remain uncoated and result as final defects in the sintered body. Surface micro-porosity appears more abundant on high solid loading samples. Such microstructure is very promising from a biological perspective because the surface micro-porosity can provide grips for phyllopods of bone cells and, together with macro-porosity, act as channels for the diffusion of nutrients.

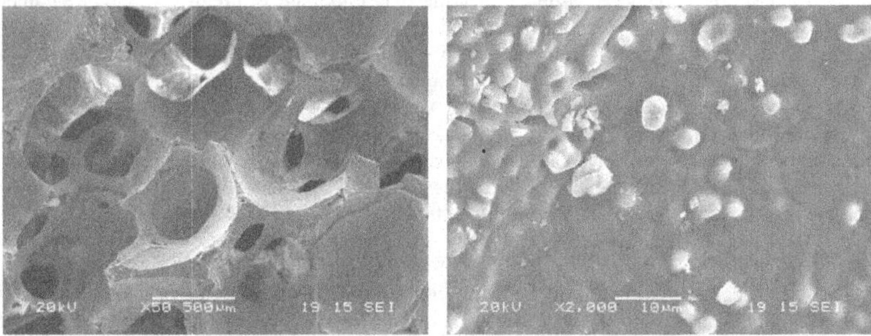

Figure 2. SEM micrographs of porous HA scaffolds produced with low solid loading.

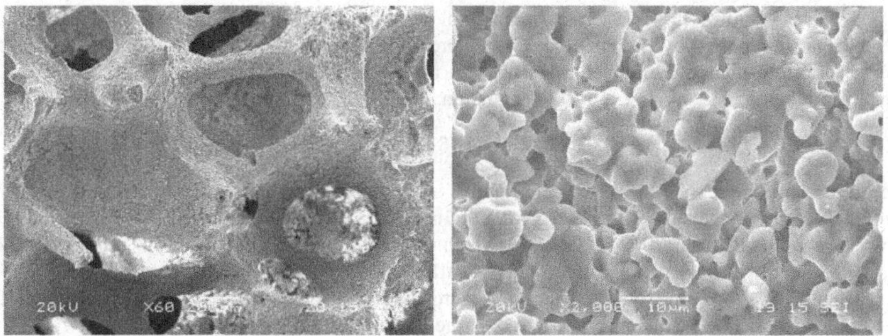

Figure 3. SEM micrographs of porous HA scaffolds produced with high solid loading.

Table I shows the results in terms of pore size, porosity and mechanical properties. The porosity of the samples was measured indirectly from the bulk density and by using a density for pure HA equal to 3.156 g/cm^3. The compressive strength was obtained by dividing the maximum force (*i.e.* the load leading to the collapse of the sintered sample) by the theoretical cross-section. As expected the average pore size and total porosity decrease as the solid loading increases while the compression resistance increases.

Table I. Pore size, porosity and compressive strength.

Solid loading	Average pore size [μm]	Porosity [%]	Strength [MPa]
Low	510	88	0,3±0,1
Medium	445	80	0,9±0,5
High	390	73	2,2±0,5

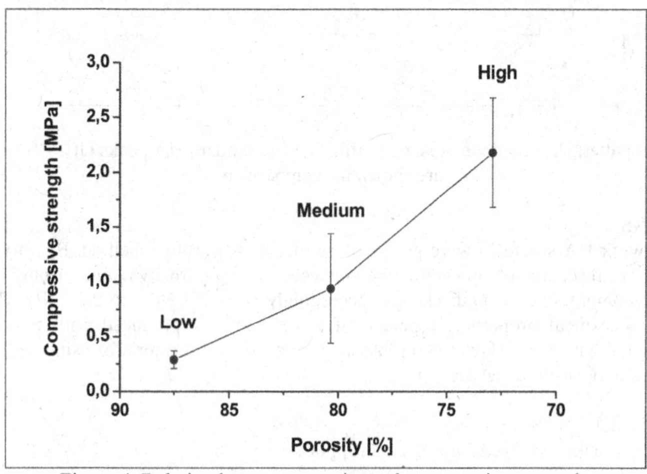

Figure 4. Relation between porosity and compressive strength.

Figure 4 shows the relation between total porosity and compressive strength. The trend resembles those reported by Bignon et al. [5] in a previous paper though the porosity range considered here is definitely higher. Moreover, Macchetta et al. [6] in previous work related the compressive strength and the porosity by an exponential equation:

$$y = 294.18\ e^{-6.71x} \tag{1}$$

where y is the compressive strength (MPa) and x is the porosity volume fraction (from 0 to 1). An increase of compressive strength from 2.3 MPa to 36.4 MPa was calculated for HA/TCP scaffold for a decrease of porosity from 70% to 30%. Such values are comparable with data in Table I for 72% porosity and could predict a similar trend in HA scaffolds produced in the present work.

Figure 5 shows the XRD spectrum recorded on the powder obtained after crushing the produced scaffolds. One can appreciate that only the fundamental HA peaks are present.

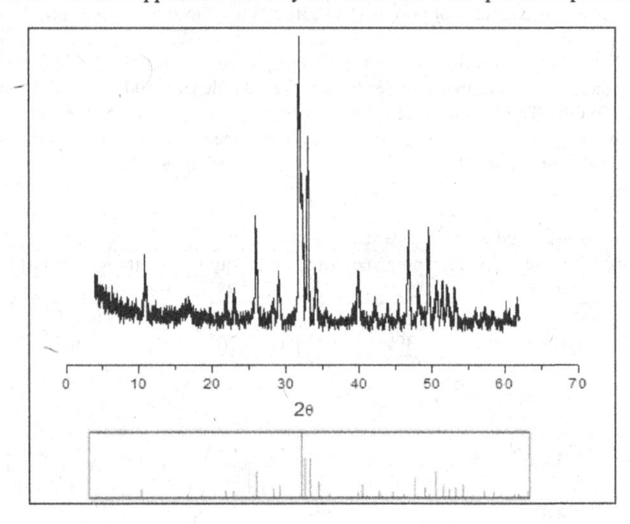

Figure 5. XRD pattern recorded the crushed scaffolds; the standard HA peaks (JCPDS card n°09-432) are shown for comparison.

CONCLUSIONS

In this work HA scaffolds were produced by the sponge-replica method. By varying the solid loading during the impregnation procedure interconnected macro-porosity ranging from 70 to 85% was obtained. The compressive strength changes accordingly from 0.3 MPa to 2.2 MPa. The obtained physical and mechanical properties, especially for the scaffolds produced with the highest solid loading, associated with the surface micro-porosity, are positive indicators for using the manufactures porous ceramics in bone-tissue engineering.

ACKNOWLEDGEMENTS

Eurocoating Spa is acknowledged for financial support.

REFERENCES

[1]B.S. Chang et al., Osteoconduction at porous hydroxyapatite with various pore configurations, *Biomaterials,* **21**(12), 1291–8 (2000).
[2]J.X. Lu et al., Role of interconnections in porous bioceramics on bone recolonization in vitro and in vivo, *J. Mater. Sci. Mater. Med.,* **10**(2), 111–20 (1999).
[3]M. Milosveski, Preparation and properties of dense and porous calcium phosphate, *Ceramics International*, **25**(8), 693-696 (1999).
[4]S. Teixeira et al., Proliferation and mineralization of bone marrow cells cultured on macroporous hydroxyapatite scaffolds functionalized with collagen type I for bone tissue regeneration, *J. Biom. Mat. Res.*, **95A**, 1–8 (2010),
[5]A. Bignon et al., Effect of micro and macroporosity of bone substitutes in their mechanical properties and cellular response, *J. Mater. Sci. Mater. Med.*, **14,** 1089-97 (2003).
[6]A. Macchetta et al. Fabrication of HA/TCP scaffolds with a graded and porous structure using a camphene-based freeze-casting method, *Adv. Biomat.*, **5**, 1319–1327 (2009).

HOLLOW HYDROXYAPATITE MICROSPHERES FOR CONTROLLED DELIVERY OF PROTEINS

H. Fu, M. N. Rahaman, and D. E. Day
Department of Materials Sciences and Engineering, and Center for Bone and Tissue Repair and Regeneration, Missouri University of Science and Technology, Rolla, MO 65409, USA

ABSTRACT

Hollow hydroxyapatite (HA) microspheres were prepared by reacting solid microspheres of a lithium–calcium–borate glass (106–150 microns) for 2 days in K_2HPO_4 solution (0.02 M; 37°C), and evaluated as a controlled delivery device for a model protein, bovine serum albumin (BSA). The as-prepared HA microspheres had a hollow core with a diameter equal to 0.6 the external diameter, a surface area of $\surd$100 m^2/g, and a mesoporous shell wall (pore size of $\approx$13 nm). After loading the hollow HA microspheres with a solution of BSA, release of the BSA into a medium of phosphate-buffered saline (PBS) was measured as a function of time using a micro bicinchoninic acid (BCA) protein assay reagent. BSA release initially increased linearly with time, but almost ceased after 24–48 hours. Modification of the BSA release kinetics was achieved by altering the microstructure of the shell wall of the as-prepared HA microspheres using a controlled heat treatment (5 h at 600°C–900°C). Sustained release of BSA was achieved over than 7–14 days from HA microspheres heated for 5 h at 600°C. The potential application of these hollow HA microspheres as a device for controlled local delivery of protein growth factors and drugs is discussed.

1. INTRODUCTION

Proteins such as growth factors and drugs have become an important class of therapeutic agents for tissue healing and regeneration. Some of these compounds have short *in vivo* half-lives, so they need to be administered repeatedly, sometimes by invasive methods such as subcutaneous injection, resulting in pain and inconvenience to the patient. Moreover, some compounds may show toxic effects, requiring site specific delivery to reduce toxic burden to other parts of the body. A key requirement for the successful use of growth factors and drugs in therapeutic applications is a controlled delivery system that allows the proteins to be delivered locally to the target site at concentrations within the therapeutic limits and for the required duration [1].

Natural and synthetic biodegradable polymers have found wide application as carrier materials for protein delivery [2]. The delivery systems include microspheres, hydrogels, and three-dimensional porous scaffolds [3,4]. These polymers degrade *in vivo*, either enzymatically or non-enzymatically, to produce biocompatible or non-toxic by-products along with progressive release of the dispersed or dissolved protein. Natural polymers and their derivatives in the form of gels or sponges have been used extensively as delivery vehicles. In particular, collagen is a readily available extracellular matrix component that allows cell infiltration and remodeling, making it an attractive delivery system for proteins [5,6]. Biodegradable synthetic polymers, such as poly(lactic acid), PLA, and poly(glycolic acid), PGA, as well as their copolymers, poly(lactic co-glycolic acid), PLGA, are the most widely-used delivery systems. In addition to being widely available, they can be prepared with well-controlled, reproducible chemical and physical properties [2–4,7]. They are also among the few synthetic biodegradable polymers approved by the Food and Drug Administration (FDA) for *in vivo* use.

Inorganic materials that have been utilized as carriers for protein delivery consist primarily of calcium phosphate materials such as β-tricalcium phosphate, $Ca_3(PO_4)_2$ and hydroxyapatite, HA, $Ca_{10}(PO_4)_6(OH)_2$ [1], and bioinert metal oxides such as silica, SiO_2 [8,9]. The calcium phosphate materials, composed of the same elements as bone, are biocompatible and produce no systemic toxicity or immunological reactions. In addition to its chemical and structural stability, SiO_2 can be prepared near room temperature by sol-gel methods and other solution-based methods, so the protein activity can be retained. The inorganic delivery systems typically consist of nanoparticles, porous particles, granules, or porous substrates in which the protein is adsorbed or attached to the surfaces of the porous material, or encapsulated within the pores [8–11].

Day and Conzone [13] invented a process for preparing porous phosphate materials with high surface area by converting borate glasses with special compositions in an aqueous phosphate solution near room temperature [14,15]. A characteristic feature of the process is that it is pseudomorphic, so the HA product retains the same external shape and dimensions of the starting glass object. Using this process, Wang et al. [16] reacted solid glass microspheres (106–125 μm) in 0.25 M K_2HPO_4 solution. They found that the product consisted of hollow microspheres of a calcium phosphate material which, on heating for 4 h at 600°C, converted to HA. Huang et al. [17] prepared hollow HA microspheres by reacting glass microspheres under similar conditions used by Wang et al. [16]. They measured the surface area (135 m^2/g) of the smaller HA microspheres (106–125 μm) and the rupture strength (1.6 MPa) of the larger HA microspheres (500-800 μm), and studied the effect of heat treatment on the surface area and rupture strength.

Our previous work showed that the microstructure of hollow HA microspheres prepared by converting Li_2O–CaO–B_2O_3 glass microspheres (106–150 μm) in a K_2HPO_4 solution can be modified over a wide range by controlling the process variables [18]. By varying the concentration (0.01–0.25 M) and the temperature (25–60°C) of the K_2HPO_4 solution at pH = 9–12, hollow HA microspheres with a hollow core diameter to microspheres diameter (d/D) of 0.14–0.62, surface area of 78–145 m^2/g, and pore size of 8–20 nm were produced.

The objective of this work was to evaluate hollow HA microspheres prepared by the glass conversion process as a potential device for controlled delivery of proteins. Bovine serum albumin (BSA) was used as a model protein in this work because it is one of the most widely studied proteins. The ability to fill the hollow HA microspheres with an aqueous solution of BSA, and the release kinetics of BSA from the filled microspheres into an aqueous medium were determined. Modification of the BSA release kinetics was studied by altering the microstructure of the shell wall of the as-prepared HA microspheres using a controlled heat treatment (5 h at 600°C or 900°C).

2. EXPERIMENTAL PROCEDURE

2.1 Preparation of hollow hydroxyapatite (HA) microspheres

Hollow HA microspheres were prepared by reacting solid glass microspheres in an aqueous phosphate solution as described previously [18]. Briefly, borate glass, with the composition (wt%): 15CaO, 10.63Li_2O, 74.37 B_2O_3, designated CaLB3-15, was prepared by melting Reagent grade $CaCO_3$, Li_2CO_3 and H_3BO_3 (Alfa Aesar, Haverhill, MA, USA) in a Pt crucible at 1200ºC for 45 min, and quenching the melt between cold stainless steel plates. Particles of size 106–150 m were obtained by grinding the glass in a hardened steel mortar and pestle, and sieving through 100 and 140 mesh sieves. Microspheres were obtained by dropping the crushed particles down a vertical tube furnace at 1000°C, as described in detail elsewhere [14].

Hollow HA microspheres were prepared by reacting the solid glass microspheres for 2 days in 0.02 M K_2HPO_4 solution at 37°C and pH = 9.0. These conditions were used because our previous work showed that they resulted in the formation of hollow HA microspheres with large d/D value (ratio of the hollow core diameter to the sphere diameter), high surface area, and mesoporous shell wall [18]. In all the experiments, 1 g of glass microspheres was placed in 200 ml solution, and the system was gently stirred continuously. Upon completion of the conversion process, the HA microspheres were washed 3 times with distilled water, soaked in anhydrous ethanol to displace residual water, dried for at least 12 h at room temperature, then for at least 12 h at 90°C, and stored in a desiccator. These 'as-prepared' HA microspheres were subjected to a controlled heat treatment to modify the microstructure of the shell wall. Microspheres were heated in a Pt crucible for 5h at temperatures of 600°C or 900°C. These heat treatment temperatures were used because they were below and above the onset temperature for sintering (densification) of the porous HA shell.

2.2 Characterization of hollow HA microspheres

The microstructure of the external surface and the cross section of the as-prepared and heat-treated HA microspheres was examined in a scanning electron microscope, SEM (S-4700; Hitachi, Tokyo, Japan), at an accelerating voltage of 10 kV and working distance of 12 mm. The specific surface area and the pore size distribution of the shell wall of the HA microspheres were measured using nitrogen adsorption (Autosorb-1; Quantachrome, Boynton Beach, FL). Prior to the measurement, a known mass of microspheres (in the range 300–500 mg) was weighed, and evacuated for 15 h at 120°C to remove adsorbed moisture. The volume of nitrogen adsorbed and desorbed at different relative gas pressures was measured and used to construct adsorption–desorption isotherms. The first five points of the adsorption isotherm, which initially followed a linear trend implying monolayer formation of adsorbate, were fitted to the Brunauer–Emmett–Teller (BET) equation for the determination of the specific surface area [19]. The pore size distribution was calculated using the Barrett–Joiner–Halenda (BJH) method applied to the desorption isotherms [20].

The rupture strength of individual HA microspheres, as-prepared or heat-treated, was measured in a nano-mechanical testing machine (Nano Bionix; Agilent Technologies; Oak Ridge, TN) using a procedure described in detail elsewhere [17]. Because of the difficulty of manipulating microspheres of size 106–150 μm in the testing machine, larger spheres (diameter 200–250 μm) were tested. At least 8 microspheres were measured for each group (as-prepared or heat-treated), and the rupture strength was expressed as a mean value ± standard deviation.

2.3 Biocompatibility of hollow HA microspheres

The biocompatibility of the hollow HA microspheres was evaluated by examining their ability to support cell proliferation in vitro. For these experiments, thin discs of hollow HA microspheres were formed by pouring the borate glass microspheres into a graphite die, heating the system for 1 h at 550°C to join the glass microspheres, and then reacting the disc in 0.02M K_2HPO_4 solution to convert the glass microspheres to hollow microspheres. After sterilization by washing 3 times with water and ethanol, followed by heating for at least 24 h at 120°C, the discs composed of hollow HA microspheres were seeded with 60,000 MC3T3-E1 cells suspended in 60 μl medium, and incubated for 4 h to permit cell attachment. The cell-seeded constructs were then transferred to a 24-well culture plate containing 2 ml of complete medium per well. All cell cultures were maintained at 37°C in a humidified atmosphere of 5% CO_2 with the medium changed every 2 days. At selected time intervals, disks with attached cells were removed, washed, dehydrated and examined in the SEM (S-4700; Hitachi).

Total protein in lysates recovered from the cell-seeded discs composed of hollow HA microspheres was measured with a micro-BCA Protein Assay Kit (Piece Biotechnology, Rockford, IL) to assess cell proliferation on the discs. Cells were detached from the spheres discs by lysis in 500 μl of 1% triton solution for 1h. Aliquots of the released lysate were mixed with 100 μl of micro-BCA working reagent and the resultant mixture incubated at 37°C for 2 h. The absorbance of each solution was measured at 550 ηm using a HP 8452A diode array spectrophotometer (BMG LABTECH Inc., Cary, NC), with bovine serum albumin (BSA) used as a standard for comparison.

2.4 Loading of BSA into hollow HA microspheres

The ability to load a model protein, bovine serum albumin (BSA), into the hollow HA microspheres and the distribution of the BSA in the microspheres was studied using optical microscopy. To permit visual observation of the protein distribution within the microspheres, a fluorescein isothiocyanate-labeled BSA (referred to as FITC–BSA) (Sigma–Aldridge, St. Louis, MO) was used. A mass of 0.1 g microspheres was immersed in 5 ml of a solution consisting FITC–BSA in phosphate-buffered saline (PBS) (FITC–BSA concentration = 5 mg/ml). A low vacuum was applied to the system to remove air trapped within the microspheres, thereby assisting the incorporation of the FITC–BSA into the microspheres. When the removal of air bubbles from the microspheres had ceased (as determined visually), the microspheres loaded with FITC–BSA were dried in air at room temperature, and observed in an optical microscope.

2.5 Release kinetics of BSA from hollow HA microspheres into PBS

The hollow HA microspheres were loaded with a solution of BSA (without FITC labeling) using the method outlined above, removed, and immersed in PBS to determine the release of BSA from the microspheres into the PBS. In the loading step, 200 mg of the microspheres was placed in 2 ml of a PBS solution containing 5 mg/ml BSA (molecular weight = 66 kDa; Sigma–Aldrich). The BSA-loaded microspheres were removed, rinsed three times with PBS, and placed in a beaker containing 20 ml PBS. The system was kept at 37°C, and the solution was stirred continuously. At selected time intervals, 50 μL aliquots were taken from the solution and used for determining the amount of BSA released into the solution.

The concentration of BSA in each aliquot was measured using a micro bicinchoninic acid (BCA) protein assay reagent (Product # 23240ZZ; Thermo Fisher Scientific, Rockford, IL). This assay is very sensitive to dilute concentrations of proteins and has a linear working range of 0.5–20 μg/ml for BSA [21]. The aliquots were mixed with 50 μl PBS and 100 μl of the working reagent, reacted for 2 h at 37°C and cooled to room temperature. The absorbance of each solution was measured at 550 nm using a HP 8452A diode array spectrophotometer (BMG LABTECH Inc., Cary, NC). The concentration of BSA was determined from a standard curve calibrated from measurements of the absorbance of PBS containing known concentrations of BSA. At the completion of the BSA release experiments, the microspheres were washed 3 times with distilled water and dried in air at room temperature. The dried microspheres were crushed and the residual BSA in the microspheres was measured using the BCA technique described above.

3. RESULTS

3.1 Microstructure and strength of as-prepared and heat-treated HA microspheres

Figure 1 shows an optical image of the starting CaLB3-15 glass microspheres (Fig. 1a), and SEM images showing the external surface and the cross-section of a hollow HA microsphere formed by reacting the glass microspheres for 2 days in 0.02 M K_2HPO_4 solution (pH = 9) at 37°C (Figs. 1b, 1c). The HA microspheres had a hollow core diameter, relative to the external diameter of the hollow microspheres, $d/D = 0.61 \pm 0.03$, surface area = 101 ± 5 m^2/g, and a pore size of the shell wall = 13 ± 2 nm. XRD, FTIR, and EDS analysis (not shown) confirmed that the hollow microspheres had a structure and composition corresponding to that of a hydroxyapatite (HA)-type material.

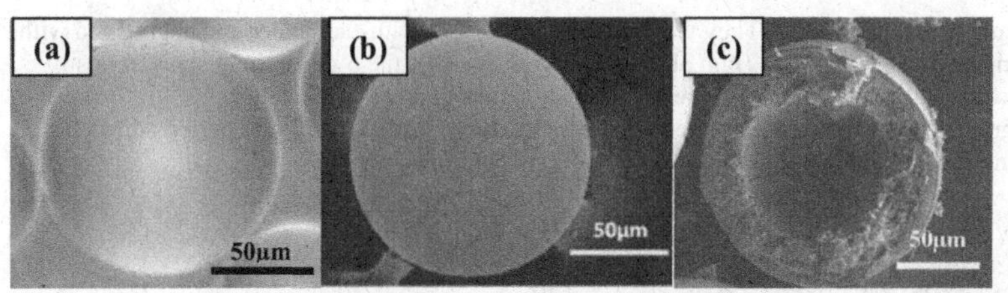

Fig. 1. Optical and SEM images of (a) starting glass (CaLB3-15) microspheres, (b) external surface of a hollow HA microsphere prepared by converting the glass microspheres for 48 h in 0.02M K_2HPO_4 solution at 37°C and pH = 9, (c) cross section of hollow HA microsphere.

SEM images of the surface and cross section of the hollow HA microspheres, as-prepared and after heat treatment for 5 h at 600°C or 900°C, are shown in Fig. 2. As-prepared, the surface consisted of a mesoporous structure of fine, plate-like (or needle-like) HA particles (Fig. 2a). The cross section (Fig. 2d) shows that the shell wall consisted of two distinct layers: a less porous surface layer of thickness \/5 µm and a more porous inner layer. Heating for 5 h at 600°C did not produce a measurable change in the porosity of the hollow HA microspheres, but resulted in a marked change in the surface microstructure (Fig. 2b). The particles in the surface layer had a more rounded morphology, with a size <50 nm. Except for coarsening, little change in the microstructure of the inner layer of the shell wall was observed (Fig. 2e). After heating for 5 h at 900°C, fine pores remained on the surface of the microspheres (Fig. 2c), but the cross section of the shell wall was almost fully dense (Fig. 2f).

The rupture strength of the as-prepared hollow HA microspheres (200–250 µm) was 11 ± 6 MPa. After heating for 5 h at 600°C, the rupture strength increased to 17 ± 8 MPa, while heating for 5 h at 900°C resulted in a further increase in the rupture strength (30 ± 10 MPa).

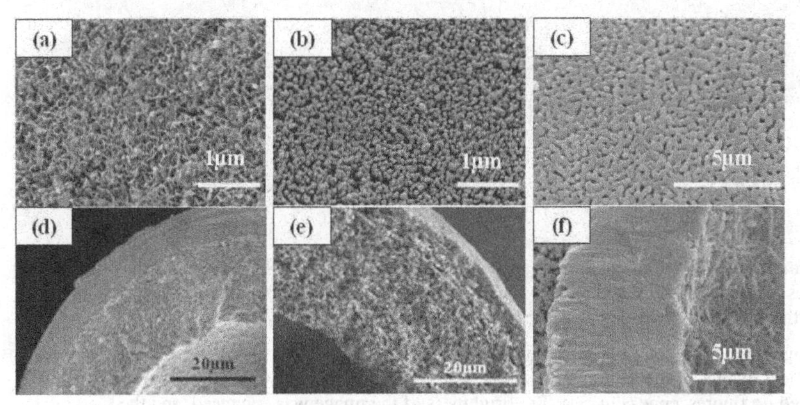

Fig. 2. SEM images of the surface (a)–(c) and cross section (d)–(f) of the shell wall of hollow HA microspheres: (a, d) as-prepared; (b, e) heated for 5 h at 600°C; (c, f) heated for 5 h at 900°C.

3.2 Biocompatibility of hollow HA microspheres

Figure 3 shows SEM images of MC3T3-E1 cells on hollow HA microspheres after incubation for 2, 4, and 6 days. The cells seen in the micrographs appeared to attach to the HA microspheres by day 2, and increased in density with the culture duration (Figs. 3b, 3c). After 4 days, the cells tried to colonize the surface of the spheres. They were in physical contact with each other and aggregated with the neighboring cells via extensions (Fig. 3b). The SEM images for the 6-day culture show almost complete coverage of the scaffolds with the MC3T3-E1 cells and increased cell density (Fig. 3c). Viewed as a group, the continuous increase in cell density during the 6 day culture period shows the ability of the hollow HA microspheres to support cell proliferation.

Fig. 3. SEM images of MC3T3-E1 cell morphology on hollow HA microsphere after culturing for 2 (a), 4 (b) and 6 (c) days.

Results of the quantitative assay of total protein in cell lysates recovered from the hollow HA microspheres after incubation for 2, 4, and 6 days are shown in Fig. 4. The amount of protein recovered from the scaffolds showed a nearly linear increase in cell proliferation during the 6 day incubation, a finding that complements the observations from the SEM images.

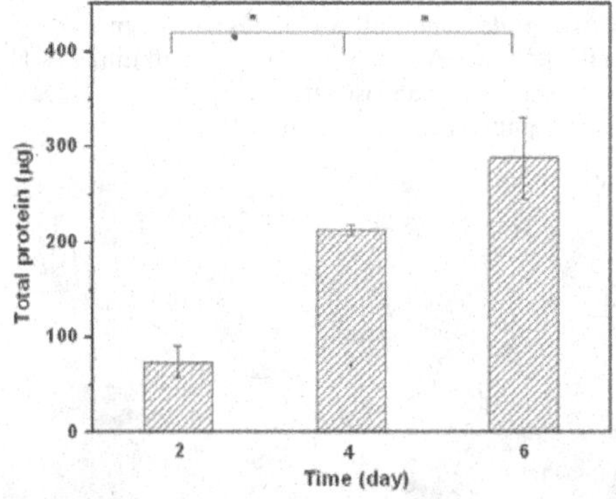

Fig. 4. Quantitative measurement of total protein content per disc of HA microspheres in MC3T3-E1 cell cultures incubated for 2, 4 and 6 days. Mean±sd; n=4. *Significant increase in the total amount of protein on the discs with increasing culture duration ($p<0.05$)

3.4 Loading and distribution of BSA in hollow HA microspheres

Optical images of the as-prepared hollow HA microspheres prior to loading with FITC-labeled BSA showed no fluorescence (Fig. 5a). The brightness of the image was enhanced, and the microspheres were circled to reveal their presence. In comparison, the surfaces of the microspheres filled with the FITC-labeled BSA solution showed a high degree of fluorescence (green color), indicating the presence of BSA (Fig. 5b). In order to show the distribution of the BSA, after filling with the BSA solution and drying, the microspheres were sectioned and observed in the microscope. The fluorescence in Fig. 5b (inset) shows that the BSA was incorporated within the hollow core of the microspheres as well as within the mesoporous shell wall.

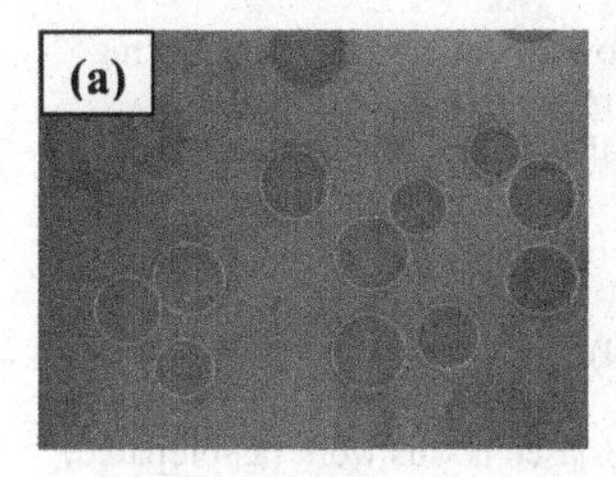

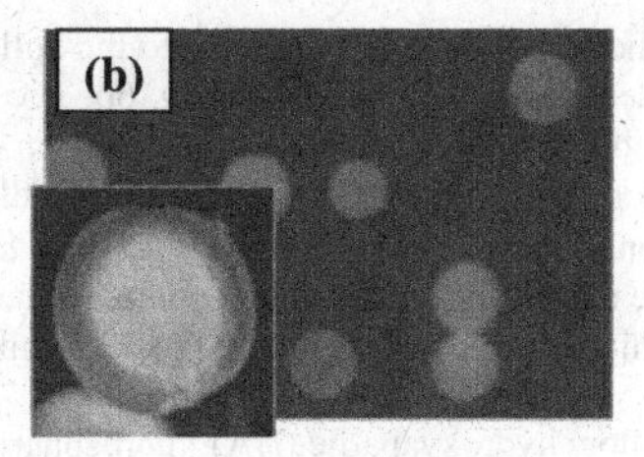

Fig. 5. Optical images of the surface of (a) as-prepared HA microspheres (brightness enhanced to show microspheres (circles)), and (b) HA microspheres loaded with fluorescent FITC-labeled BSA. (Inset: cross section of microspheres loaded with FITC-labeled BSA.)

3.5 Release kinetics of BSA from hollow HA microspheres

Figure 6 shows the BSA release kinetics from the as-prepared and heat-treated HA microspheres into a PBS solution. Each group of microspheres was previously loaded with a solution consisting of 5 mg of BSA per ml of PBS. Release of BSA from the as-prepared HA microspheres was initially rapid (~2.0 µg/ml/h) during the first 10–12 h, then slowed considerably, with 95% of the final amount released within 24 hours, and almost ceased after 24–48 h (Fig. 6a). The total amount of BSA released into the PBS after 14 days was √22 µg/ml, which was √44% of the amount of BSA initially loaded into the microspheres (Table I). Heating the as-prepared microspheres for 5 h at 600°C resulted in a marked increase in the amount and duration of the BSA released into the PBS. The amount of BSA released was lower than that for the as-prepared microspheres during the first 48–72 h, but continued to increase, with 95% of final amount released in 7 days, and almost ceased after 14 days. The total amount of BSA released (after 14 days) was √35 µg/ml, which was √30% of the BSA initially loaded into microspheres. For the HA microspheres heated for 5 h at 900°C, release of the BSA from the microspheres into the PBS was limited. The cumulative amount of BSA released after 3–5 h was √2 µg/ml, and it remained at this value thereafter.

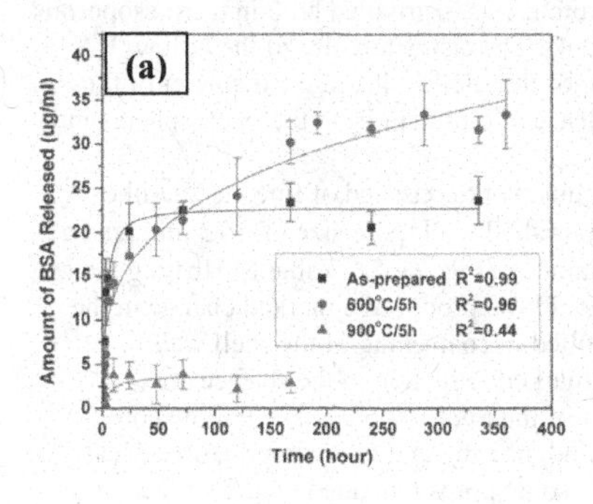

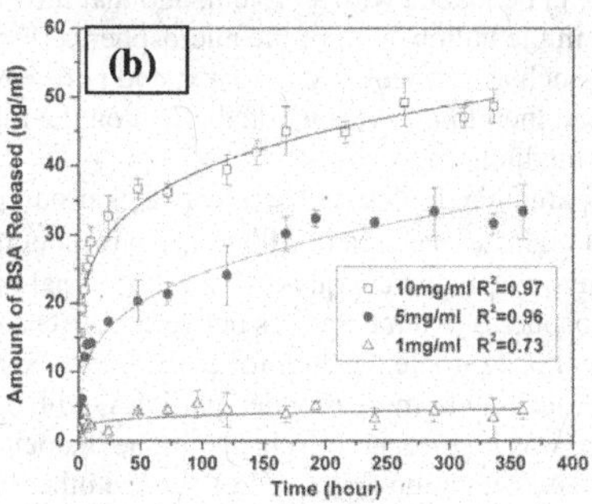

Fig. 6. Amount of BSA released from hollow HA microspheres into PBS: (a) as-prepared HA microspheres, and after heat treatment under the conditions shown. (The microspheres were loaded with a solution of 5 mg BSA per ml PBS prior to immersion in the PBS). (b) HA microspheres heat treated at 600°C for 5h, for different concentrations of BSA loaded into the microspheres. (The R^2 value for each fitted curve is also shown.)

The effect of varying the amount of BSA loaded into the hollow HA microspheres (1–10 mg BSA per ml of PBS) on the release kinetics is shown in Fig. 6b. The microspheres used in these experiments were heated for 5 h at 600°C. For a BSA concentration of 1 mg/ml, the total amount of BSA released was limited (√5 μg/ml). Higher BSA loading markedly enhanced the amount of BSA released, as well as the duration of the release. The release kinetics from microspheres loaded with 10 mg/ml BSA followed the same trend as those for microspheres loaded with 5 mg/ml BSA, but the total amount of BSA released was higher (45 μg/ml compared to 35 μg/ml).

Table I. Characteristics of hollow hydroxyapatite (HA) microspheres used in this work (as-prepared, and heat treated under the conditions shown), and summary of BSA release from the hollow HA microspheres into a medium of PBS.

Sample	Surface area (m^2/g)	Rupture strength (MPa)	BSA loading (mg BSA/g HA)	Total BSA released (mg BSA/g HA)	Duration of release (days)
As-prepared	102	11 ± 6	50	22	1–2
600°C/5 h	19	20 ± 10	119	35	7–14
900°C/5 h	2	30 ± 10	<5	2	<1

4. DISCUSSION

The hollow HA microspheres evaluated in this work provide a promising inorganic device for the controlled delivery of proteins. These hollow HA microspheres, with a composition similar to the main mineral constituent of human bone, are biocompatible. By modifying the microstructure of the shell wall of the microspheres using a controlled heat treatment, and by varying the concentration of the protein initially loaded into the microspheres, both the amount of protein released from the microspheres and the duration of the release can be varied.

Using a model protein with a fluorescent label (FITC–BSA), it was shown that the hollow HA microspheres can be loaded with a protein, and that the protein was distributed both in the mesoporous shell wall and in the hollow core of the microspheres (Fig. 4). The characteristics of the hollow HA microspheres, such as surface area, porosity, and pore size of the shell wall and the diameter of the hollow core, can therefore have a marked effect on the release of a protein from the microspheres into a surrounding medium.

As prepared, the hollow HA microspheres used in this work consisted of fine, needle-like crystals with a high surface area (√100 m^2/g), a mesoporous shell wall (pore size of ≈13 nm), and a hollow core having a diameter equal to 0.6 the external diameter (Fig. 2a, 2d; Table I). Heating the as-prepared microspheres at temperatures between 600°C and 900°C produced a marked change in the microstructure. For example, heating for 5 h at 600°C resulted in coarsening of the shell wall microstructure, leading to more rounded particles, with a marked reduction in the surface area. However, there was no measurable change in the diameter of the microsphere or the hollow core, indicating little densification (Fig. 2b, 2e). On the other hand, heating at higher temperature (5 h at 900°C) resulted in densification of the shell wall, although some pores remained on the surface of the microsphere (Fig. 2c, 2f).

Modification of the shell wall microstructure by heat treatment provided the ability to markedly influence the release of BSA from the microspheres. Both the amount of BSA released as well as the duration of the release was influenced by the heat treatment. The release of BSA from the as-prepared HA microspheres was rapid, and essentially ceased after 24–48 h (Fig. 6). The cumulative amount of BSA released in the medium after 14 days was 22 μg/ml, which was 44% of the BSA initially loaded into the microspheres. Presumably, much of the BSA loaded into the microspheres was adsorbed on to the high surface area of the shell wall, and released rapidly upon immersion of the BSA-loaded microspheres into the PBS. Little release of BSA was found for the HA microspheres heated for 5 h at 900°C. SEM observation showed that the shell wall for this group of HA microspheres was dense (Fig. 2f), so presumably no BSA was loaded into the microspheres. The little BSA released was presumably due to BSA adsorbed on the surface of the microspheres. When compared to the as-prepared HA microspheres, a far greater amount of BSA was loaded into the microspheres heat treated for 5 h at 600°C (Table I), and presumably much of the BSA was contained within the hollow core. Upon immersion of the BSA-loaded microspheres into the PBS, the initial rapid release presumably resulted from the BSA within the shell, while the more sustained release resulted from diffusion of the BSA from the core through the shell wall.

The amount of BSA released from the hollow HA microspheres can also be varied by changing the concentration of BSA initially loaded into the microspheres (Fig. 6(b)). At any time, the cumulative amount of BSA released into the PBS increased with the amount of BSA initially loaded into the microspheres. Furthermore for BSA concentrations of 5μg/ml and 10μg/ml loaded into the microspheres, the release kinetics showed the same trend, indicating that the mechanism of BSA release was the same for these two concentrations.

The results showed that only 30–40% of the BSA initially incorporated into the HA microspheres was released into the PBS (Table I). The factors that limit the release of larger amounts of BSA into the PBS are not clear. However, it should be recalled that the loading of the BSA solution into the hollow HA microspheres was assisted by a small pressure gradient, resulting from the application of a small vacuum to the system to remove the air from within the hollow HA microspheres. On the other hand, the release of the BSA into the PBS was driven by the concentration gradient of BSA between the HA microspheres and the PBS. As the concentration gradient decreased with time, the release of BSA became slower, and eventually ceased.

CONCLUSION

Hollow hydroxyapatite (HA) microspheres (106–150 μm) with a hollow core diameter equal to 0.6 the external diameter and a mesoporous shell wall were prepared by a low temperature glass conversion route and evaluated as a device for controlled delivery of a model protein, bovine serum albumin (BSA). Both the hollow core and the mesopores of the shell wall were loaded with a solution of BSA. Release of the BSA from the as-prepared HA microspheres increased linearly time, and ceased after 24–48 hours. The amount of BSA released from the microspheres and the duration of the release was varied by heat treating the as-prepared HA microspheres to modify their microstructure and by varying the amount of BSA loaded into the microspheres. For HA microspheres heated for 5 h at 600°C, 30–40% of the BSA initially loaded into the microspheres was released over 7–14 days. In general, the present results show promising potential for the application of these hollow HA microspheres as a biocompatible inorganic device for controlled local delivery of proteins such as growth factors and drugs.

ACKNOWLEDGEMENT

This work was supported by the National Institute of Dental and Craniofacial Research, National Institutes of Health, Grant # 1R15DE018251-01.

REFERENCES

[1] S. K. Mallapragada and B Narasimhan, Drug Delivery Systems, In Handbook of Biomaterials Evaluation 2nd ed., von Recum AF (Ed.), Taylor & Francis, Philadelphia, PA, 1999; pp. 425-437.

[2] V. R. Sinha and A. Trehan, Biodegradable Microspheres for Protein Delivery, J. Control Release, 90, 261-80 (2003).

[3] Y. Tabata, Tissue Regeneration Based on Growth Factor Release, Tissue Eng., 9 (Suppl. 1), S5-S15 (2003).
[4] R. R. Chen and D. J. Mooney, Polymeric Growth Factor Delivery Strategies for Tissue Engineering, Pharm. Res., 20, 1103-12 (2003).
[5] S. Ma, G. Chen, and A. H. Reddi, Collaboration between Collagenous Matrix and Osteogenin is Required for Bone Induction, Ann. NY Acad. Sci,, 580, 525-35 (1990).
[6] J. M. McPherson, The Utility of Collagen-based Vehicles in Delivery of Growth Factors for Hard and Soft Tissue Wound Repair. Clin. Mater., 9, 225-34 (1992).
[7] R. L. Cleek, K. C. Ting, S. G. Eskin SG, and A. G. Mikos, Microparticles of Poly(DL-Lactic-co-Glycolic Acid)/Poly (Ethylene Glycol) Blends for Controlled Drug Delivery, J. Control Release, 48, 259-68 (1997).
[8] I. Ono, T. Ohura, M. Murata, H. Yamaguchi, Y. Ohnuma, and Y. Kuboki, A Study on Bone Induction in Hydroxyapatite Combined with Bone Morphogenetic Protein, Plast. Reconstr. Surg., 90, 870-9 (1992).
[9] U. Ripamonti, S. Ma, B. Van den Heever, and A. H. Reddi, Osteogenin, a Bone Morphogenetic Protein, Adsorbed on Porous Hydroxyapatite Substrata, Induces Rapid Bone Differentiation in Calvarial Defects of Adult Primates, Plast. Reconstr. Surg., 90, 382-93 (1992).
[10] U. Ripamonti, Osteoinduction in Porous Hydroxyapatite Implanted in Heterotopic Sites of Different Animal Models, Biomaterials, 17, 31-5 (1996).
[11] T. Matsumoto, M. Okazaki, M. Inoue, S. Yamaguchi, T. Kusunose, T. Toyonaga, et al., Hydroxyapatite Particles as a Controlled Release Carrier of Protein, Biomaterials, 25, 3807-12 (2004).
[12] Q. Peng, L. Ming, C. X. Jiang, B. Feng, S. X. Qu, and J. Weng, Preparation and Characterization of Hydroxyapatite Microspheres with Hollow Core and Mesoporous Shell, Key Eng. Mater., 309–311, 65-8 (2006).
[13] D. E. Day and S. A. Conzone, Method for Preparing Porous Shells or Gels from Glass Particles, US Patent No. 6,358,531, March 19, 2002.
[14] D. E. Day, J. E. White, R. F. Bown, and K. D. McMenamin, Transformation of Borate Glasses into Biologically Useful Materials, Glass Technol., 44, 75-81 (2003).
[15] S. D. Conzone and D. E. Day, Preparation and Properties of Porous Microspheres Made from Borate Glass, J. Biomed. Mater. Res. Part A, 88A, 531-42 (2009).
[16] Q. Wang, W. Huang, D. Wang, B. W. Darvell, D. E. Day, and M. N. Rahaman, Preparation of Hollow Hydroxyapatite Microspheres, J. Mater. Sci.: Mater. Med., 17, 641-6 (2006).
[17] W. Huang, M. N. Rahaman, D. E. Day, and B. A. Miller, Strength of Hollow Microspheres Prepared by a Glass Conversion Process, J. Mater. Sci.: Mater. Med., 20, 123-9 (2009).
[18] H. Fu, M. N. Rahaman, and D. E. Day, Effect of Process Variables on the Microstructure of Hollow Hydroxyapatite Microspheres Prepared by a Glass Conversion Process, J. Am. Ceram. Soc. 93, 3116-23 (2010).
[19] N. J. Coleman and L. L. Hench, A Gel-derived Mesoporous Silica Reference Material for Surface Analysis by Gas Sorption, 1, Textural Features, Ceram. Int., 26, 171-8 (2000).
[20] E. P. Barrett, L. G. Joyney, and P. P. Halenda, The Determination of Pore Volume and Area Distributions in Porous Substances I: Computations from Nitrogen Isotherms, J. Am. Chem. Soc., 73, 373-80 (1951).
[21] P. K. Smith, Measurement of Protein Using Bicinchoninic Acid, Anal. Biochem., 150, 76-85 (1985).

EXPRESSION OF MINERALIZED TISSUE-ASSOCIATED PROTEINS IS HIGHLY UPREGULATED IN MC3T3-E1 OSTEOBLASTS GROWN ON A BOROSILICATE GLASS SUBSTRATE

Raina H. Jain[a]*, Jutta Y. Marzillier[a]*, Tia J. Kowal[a], Shaojie Wang[b], Himanshu Jain[b], and Matthias M. Falk[a]
[a]Department of Biological Sciences, Lehigh University, Bethlehem, PA 18015, USA.
[b]Department of Materials Science and Engineering, Lehigh University, Bethlehem, PA 18015, USA.
*These authors contributed equally to this work.

ABSTRACT

Melt- and sol-gel derived bioactive glasses (BGs) are promising bone implant materials with biocompatibility superior to currently used inert titanium and ceramic implants. Recent studies indicate that the addition of boron to BGs may further enhance bone formation. While the beneficial effects of boron on bone formation and maintenance have been recognized, it is still unclear how boron, as an ultra-trace element (<1 ppm concentration), actually stimulates bone formation. We thus tested *in vitro* by examining cell adhesion, proliferation and gene-expression whether MC3T3-E1 bone-precursor cells seeded on borosilicate cover glasses would exhibit signs of differentiation into mature bone cells. As control, cells were seeded on boron-free substrates including a soda-lime silicate glass, 45S5 Bioglass®, a sol-gel derived calcium silicate BG, and tissue culture plastic. Results indicated that MC3T3-E1-cells grown for 17 days on borosilicate cover glasses up-regulated the expression of bone-specific marker proteins more significantly than on any of the other tested boron-free substrates. Since typical borosilicate glasses are expected to dissolve only insignificantly in tissue culture media, our results suggest that either extremely low concentrations of borate ions in solution (well below ppm concentrations) are sufficient to stimulate bone cell differentiation, or more likely that bone-precursor cells are able to 'sense' and react to boron that is present in the substrate. Our results shed new light on the potential role of boron on bone cell differentiation.

INTRODUCTION

Although bone implants date back to 600 A.D., the search for an ideal bone-replacement material continues[1]. In fact, with increasing life expectancy destructive lesions of bones, whether due to disease or trauma, are becoming increasingly common. Some such conditions include benign and malignant bone tumors, bone fractures, hip failure, middle ear deafness and periodontal diseases. In all cases, the ultimate goal is to replace the defective or missing bone tissue with a functioning material that will last a patient's lifetime. Bone implant materials research was revolutionized when Hench[2] discovered Bioglass® in the 1960's. Cells were shown to adhere to this glass of composition $24.5Na_2O$-$24.5CaO$-$6P_2O_5$-$45SiO_2$ (in wt.%). Also it was shown to develop a layer of hydroxyapatite ($Ca_5(PO_4)_3(OH)$) on its surface *in vivo*, the major mineral constituent of bone, making the material bioactive. Next-generation bioactive glasses, or TAMP (tailored amorphous multi-porous) scaffolds with interconnected pores of various size that promote cell adhesion and internal scaffold colonization have now been developed, using novel melt-quench and sol-gel derived techniques.[3-7] Furthermore, bioactive glasses have been found to resorb over time in the body and are replaced by natural bone; and ions (especially silicon) leaching from bioglass have been attributed to stimulate bone precursor cells to differentiate into mature, calcified matrix secreting osteoblasts[8]. These unique characteristics of bioglasses are not shared by any other currently used bone-replacement material making them superior materials for bone implants.

The element boron exists in low-abundance in the Earth's crust, mainly as water-soluble borate minerals. Borates have low toxicity in mammals (similar to table salt), but are more toxic to arthropods. Boron is an essential micronutrient of plants (<1 ppm concentration), required primarily for maintaining the integrity of cell walls[9]. In mammals, boron was found to play a crucial role in osteogenesis and maintenance of bone[10]. Under conditions of boron deficiency, development and regeneration of bone is negatively influenced.[11-13] However, surprisingly little is known on how this ultra-trace element exerts its beneficial health effects. Boron may interact with steroid hormones, and thus is involved in the prevention of calcium loss and bone de-mineralization[14]. It has also been related to vitamin D function by stimulating growth in vitamin D deficient animals[15]. Surprisingly, no recommended levels of boron have been set by FDA for intake in humans, only upper limits (20 mg/day); and due to the lack of data in humans this limit was extrapolated from animal studies[16]. Despite the lack of functional data, scientists began to evaluate the role of boron on the differentiation of osteoblasts and the formation of bone when added to cell culture media, or bioglass implants. Although somewhat contradicting reports have been published[17,18], the beneficial effect of boron on osteoblast differentiation and bone formation seems compelling[19-22]. We thought to test a potential beneficial effect of boron *in vitro* by seeding MC3T3-E1 osteoblast precursor cells on boron-containing borosilicate coverslip glasses widely used in light microscopic applications. In control experiments, cells were seeded on boron-fee substrates including soda-lime silicate glass, 45S5 and sol-gel derived bioglasses, and tissue culture (TC) plastic. Cell adhesion, proliferation, and potential differentiation were examined on all substrates by analyzing cell morphology, quantitative proliferation, and by evaluating the expression profile of a set of bone-cell specific (RunX2, BSP1, BGLAP) and of other osteoblast-relevant marker proteins (Coll A, ALP, Cx43, GAPDH) known to be upregulated specifically during osteoblast differentiation[23, 24] using quantitative real-time polymerase chain reaction (qRT-PCR) analyses.

MATERIALS AND METHODS

Substrate Materials

The 'Deutsche Spiegelglas' (round, 12 and 18 mm diameter, 0.13-0.17 mm thickness; Carolina, Item. # 633029 and 633033) borosilicate glass was used as the main substrate of this study. It is commonly used as microscope coverslips in biological studies. Its composition was analyzed using a high resolution X-ray photoelectron spectrometer (Scienta ESCA 300). To obtain overall bulk concentration of constituent oxides, the sample was fractured inside the ultra-high vacuum of the spectrometer and analyzed without exposure to ambient. In addition, to compare its effect on bone forming cells, four different boron-free substrates were used: classic 45S5 Bioglass®, a sol-gel derived $30CaO\text{-}70SiO_2$ (mol%) glass, commercial soda lime silicate glass slides (VWR), and tissue culture plastic (Falcon). The 45S5 Bioglass® was prepared by the standard melt-quench method as described elsewhere,[25] with batch composition: $24.4Na_2O\text{–}26.9CaO\text{–}2.6P_2O_5\text{–}46.1SiO_2$ (mol%). The procedure for making calcium silicate glass by the sol-gel method is given in reference 4.

Cell Lines, Culture Conditions, and Morphological Analyses

MC3T3-E1 subclone E-4 newborn mouse calvarial bone pre-osteoblasts (ATCC CRL-2593) were cultured under standard conditions in alpha-Modified Eagle's Medium (α-MEM) supplemented with 10% fetal bovine serum (FBS), 1% L-glutamine, and 100 U/ml penicillin/streptomycin in a cell culture incubator at 37°C, 5% CO_2-atmosphere, and 100% humidity as recommended by the distributor. Cells were subcultured at a ratio of 1:10 once or twice a week on regular tissue culture (TC) plastic when cells reached confluency. For all analyses cells at a low passage number (<20) were seeded on 'Deutsche Spiegelglas' borosilicate cover glasses at a density of 10,000 cells/cm^2 that were sterilized by flaming in ethanol before being placed into 3.5 cm diameter culture dishes, or 24-well culture plates. Control cells were seeded at comparable densities on soda lime glass (microscope slides cut into 1 cm^2 pieces), discs prepared from 45S5 melt-derived bioglass[25], sol-gel derived bioglass[4-7], or tissue culture plastic. To replenish nutrients, and dilute out potentially toxic cellular waste products, one-half of the culture medium was exchanged every other day. Cells growing on borosilicate cover glasses and on TC plastic were examined and imaged 1 hour, 2 days and 5 days post seeding using an inverted microscope (Nikon TE2000), phase-contrast illumination, a 20x long-distance objective, and a SPOT RT CCD camera (Diagnostic Instruments Inc., Sterling Heights, MI).

Quantitative Cell Proliferation Analyses

Cells proliferating on 4 borosilicate cover slips each for 4, 11, and 17 days, were stained with the blue, live-cell compatible nuclear stain Hoechst 33342 (Molecular Probes/Invitrogen) for 10 minutes at 1 μg/ml (prepared from 10 mg/ml stock in water) at 37°C. Three representative images per cover glass were acquired using fluorescence illumination and a DAPI filter cube. Immediately after imaging (within 15 minutes after staining) cells were processed for qRT-PCR analyses. Cell nuclei on the acquired images were counted, average cell counts (including standard deviations for the parallel cover glasses) calculated, and graphed over time using Microsoft Excel software.

Quantitative Real-Time Polymerase Chain Reaction (qRT-PCR) Analyses

The mRNA expression profile for 6 marker proteins (see Results) was analyzed in MC3T3-E1 cells when seeded (day 0), and grown for 4, 11, and 17 days on borosilicate cover glasses. Cells were lyzed and total cellular RNAs were isolated using the SYBR Green Cells-to-Ct kit (Applied Biosystems). mRNAs were reverse-transcribed using oligo-dT primer and Reverse Transcriptase (Superscript III, Invitrogen). Triplicates for 3-4 cover slips per time point, either analyzed separately, or after pooling cell lysates were analyzed on a model 7300 Thermocycler (Applied Biosystems) using sets of custom-designed oligonucleotides that corresponded to the mRNAs encoding the relevant proteins. GAPDH was analyzed in parallel and used as expression reference. Fluorescence signal detection within 35 PCR cycles was considered significant. mRNA level fold-change of relevant proteins on days 4, 11, and 17 was normalized against GAPDH expression, compared to day 0, and plotted using Microsoft Excel software. Cells growing on boron-free substrates (described above) were analyzed similarly.

RESULTS

Cell Type and Substrate Selection

To investigate the adhesion, morphology, proliferation, and especially the potential differentiation of precursor cells into mature bone cells when grown on boron-containing substrate, MC3T3-E1 newborn mouse calvarial bone pre-osteoblast cells were chosen[26]. This cell line is used as a generic osteogenic model cell line, since it has been shown to differentiate into mature, calcified matrix secreting osteoblasts under specific favorable conditions, such as the addition of phosphate and ascorbic acid[27], bone-morphogenic protein 2 (BMP-2)[28, 29], or borate ions[19, 22] to the culture medium. German 'Deutsche Spiegelglas'-type, high-quality microscope borosilicate cover glasses were chosen as cell-culture test substrate. In-house X-ray photoelectron spectroscopy revealed a chemical composition consisting of $5.7Na_2O$-$3.7K_2O$-$3.6ZnO$-$1.9TiO_2$-$7.0B_2O_3$-$78.1SiO_2$ (mol%).

Adhesion and Morphology of MC3T3-E1 Cells Grown on Borosilicate Cover Glasses

To investigate the adhesion and morphology of MC3T3-E1 pre-osteoblasts on borosilicate cover glasses, cells were harvested by trypsinisation and seeded at low density (~50 cells/mm^2) in 3.5

cm diameter tissue culture (TC) dishes containing, or not containing borosilicate cover glasses. Cells were examined and imaged 1 hour, 2 days, and 5 days post seeding. Representative images are shown in Figure 1. Cells on both substrates adhered and proliferated well, reaching complete confluency within 5 days post seeding (Figure 1, row 3). Cell adhesion occurred slightly faster on TC plastic (Figure 1, row 1), as indicated by a more adherent, spread-out morphology 1-hour post seeding. Cells developed their typical morphology within 1-2 days post seeding, without significant visible morphological differences. However, a slightly more spread-out phenotype with larger, more pronounced lamellipodial extensions was developed on the borosilicate substrate (Figure 1, row 2).

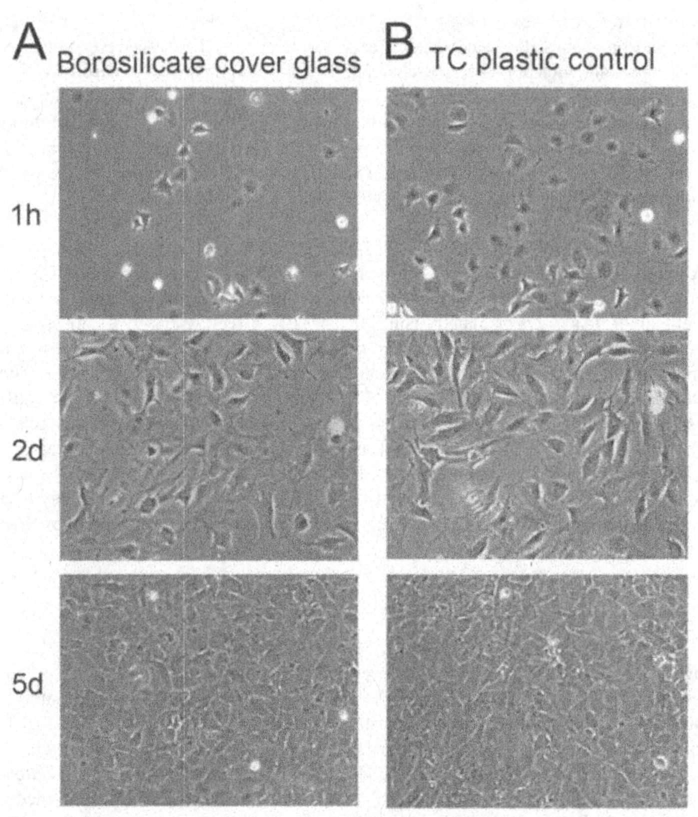

Figure 1: MC3T3-E1 osteoblast precursor cells were seeded (A) on borosilicate cover glasses and (B) on tissue culture (TC) plastic as control. Cell adhesion, morphology, and proliferation were evaluated at indicated times (see text for details).

Proliferation Rate of MC3T3-E1 Cells Grown on Borosilicate Cover Glasses

To quantitatively analyze the proliferation rate of MC3T3-E1 cells growing on borosilicate cover slips, they were stained with the blue, live-cell permeable nuclear stain Hoechst 33342 at different days post seeding. Representative areas of 3-4 cover slips for each time-point were imaged, cell nuclei counted, and cell numbers were determined and plotted over time. Representative fluorescence images of cells with cell nuclei clearly visible and acquired at days 4, 11, and 17 are shown in Figure 2A. Note the increasing numbers of cell nuclei per imaged area over time. Cell proliferation as determined from the cell-counts is depicted in Figure 2B. Quantitative analysis showed that MC3T3-E1 cells exhibited a fast proliferation rate when growing on borosilicate cover slips, approximately doubling every 30 hours, and exiting logarithmic growth by day 6-7. A comparable fast proliferation rate was determined for MC3T3-E1 cells grown on TC plastic (compare Figure 1, and data not shown).

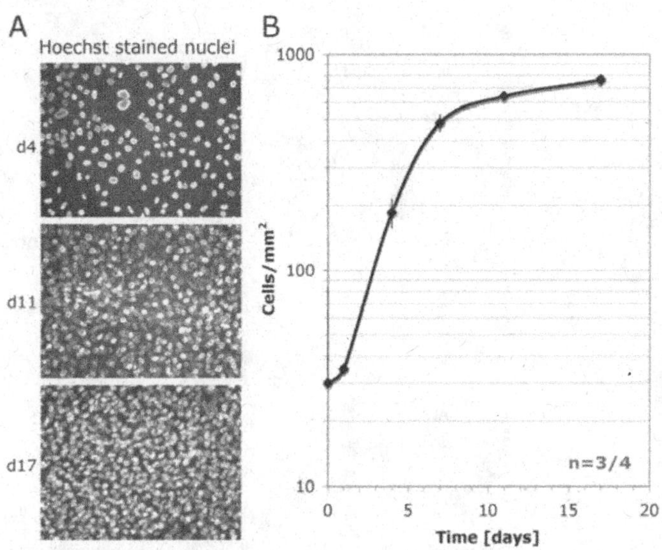

Figure 2: Proliferation of MC3T3-E1 pre-osteoblasts was determined by staining cell nuclei with the fluorescent dye, Hoechst 33342. Images were acquired with a 10x objective, nuclei counted, and cell numbers plotted over time. Representative images taken on days 4, 11, and 17 are shown in (A), a cell growth-curve extrapolated from the cell counts is shown in (B). Cells duplicated app. every 30 hours after experiencing an initial lag period, and before exiting the logarithmic growth phase at around day 6-7 (see text for further details).

Differentiation of MC3T3-E1 Cells Grown on Borosilicate Cover Glasses

To determine whether MC3T3-E1 osteoblast precursor cells would differentiate into mature, calcified matrix secreting osteoblasts when grown for longer time periods on borosilicate cover glasses, Hoechst-stained MC3T3-E1-cells were lyzed and processed for qRT-PCR analyses immediately after staining and imaging (within 15 minutes). Pilot experiments indicated that nuclear Hoechst-staining had no adverse effects on cell lysis, mRNA preparation, reverse transcription, or qRT-PCR analyses. This experimental procedure allowed for direct comparison of cell numbers determined in the proliferation assays with expression profiles detected by qRT-PCR analyses. A set of

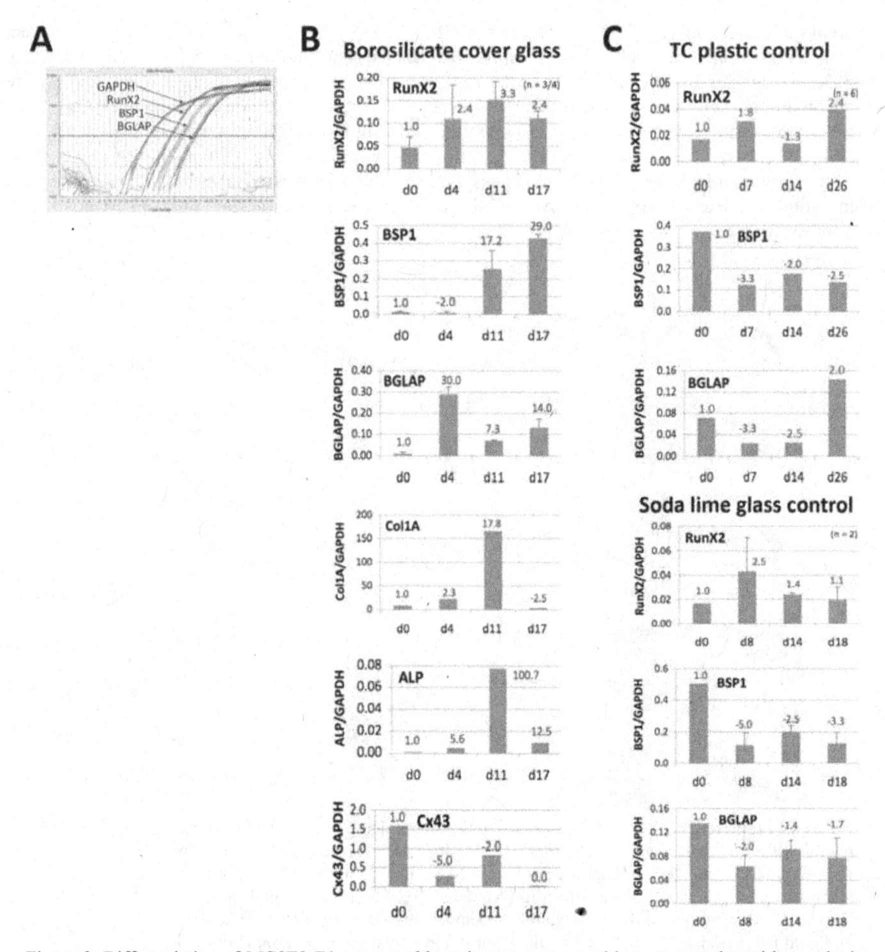

Figure 3: Differentiation of MC3T3-E1 pre-osteoblasts into mature osteoblasts was evaluated by analyzing the expression profile of 3 bone-specific (RunX2, BSP1, BGLAP), and 3 additional osteoblast-relevant proteins (Col1A, ALP, Cx43) known to be up-regulated during osteoblast differentiation by qRT-PCR analyses. (A) A representative set of obtained primary qRT-PCR fluorescence-detection curves of GAPDH, RunX2, BSP1, and BGLAP for the day of seeding (day 0) is shown. (B) Fold changes of RNA levels, normalized against GAPDH expression, for all 6 analyzed proteins of cells grown for 4, 11, and 17 days on borosilicate cover glasses compared to the expression level on day 0 (set to1) are shown (blue/grey numbers above bars). Note the strong up-regulation of most marker proteins on borosilicate cover glasses. (C) MC3T3-E1 cells grown in control on boron-free substrates (TC plastic, soda lime glass) up-regulated marker protein expression only insignificantly (see text for further details).

three bone-cell specific (RunX2 or Cbfa1, the master transcription factor responsible for bone-cell differentiation; bone sialoprotein 1/BSP1 or osteopontin/OPN; and osteocalcin or BGLAP, two bone-specific secreted proteins involved in matrix mineralization), and three other osteoblast-relevant proteins (collagen 1A/Coll A, a fibril-forming collagen abundantly expressed and secreted by bone, cornea, dermis, and tendon-forming cells; alkaline phosphatase/ALP, a membrane-bound glycosylated enzyme known to be upregulated in bone and other cell types; connexin 43/Cx43, a gap junction forming protein known to be involved in bone-cell differentiation) (for protein nomenclature see GeneCards®; http://www.genecards.org) known to be upregulated specifically during osteoblast differentiation[23,24] was analyzed and compared to the expression profile of glyceraldehyde 3-phosphate dehydrogenase/GAPDH, an important enzyme involved in carbohydrate metabolism. GAPDH was found to exhibit stable expression levels independent of cellular differentiation stages, and hence is commonly used as a housekeeping control protein.

A representative set of primary qRT-PCR fluorescence-detection curves of RunX2, BSP1, and BGLAP for the day of seeding (day 0) is shown in Figure 3A. Note the close clustering of the curves for each protein indicating highly homogenous parallel samples (n=3/4). Fold change of expression for all 6 analyzed proteins, normalized against GAPDH and compared to the expression level relative to day 0 (set to 1), is shown for days 4, 11, and 17 in Figure 3B. All proteins, except Cx43 expression levels were found to be up-regulated significantly over the 17 day long period (RunX2 = 3.3-fold; BSP1= 29.0-fold; BGLAP = 30.0-fold; Coll A = 17.8-fold; and ALP = 100.7-fold). Only Cx43 expression levels were found to be down-regulated initially by 5.0-fold before recovering by day 11; and this may be attributed to the low cell number shortly after seeding that counteracts gap junction-mediated direct cell-to-cell communication[30]. Notably, the (early) transcription factor differentiation marker, RunX2, was found to be up-regulated early (within 11 days), and less up-regulated later (day 17), while the (late) matrix protein, BSP1, was up-regulated only later (by day 17). The highest up-regulation with over 100-fold was observed for the well-recognized osteoblast differentiation marker, ALP. MC3T3-E1 cells are known to express constitutively large amounts of Coll A, and its mRNA levels were above those of GAPDH as indicated by the higher than one Coll A/GAPDH ratios (Figure 3B). All three secreted extracellular matrix proteins involved in matrix formation and mineralization (BSP1, BGLAP, and Coll A) were up-regulated significantly by 29, 30, and 18-fold (Figure 3B). For none of the boron-free substrates (soda lime silicate glass, TC plastic, melt-derived 45S5 or sol-gel derived bioglasses) a similarly significant up-regulation was observed in any of the control experiments that spanned comparable time periods (shown for RunX2, BSP1, and BGLAP of cells grown on TC plastic and soda lime glass (Figure 3C).

DISCUSSION

Taken together, our results derived from this study and presented above indicate that MC3T3-E1 bone precursor cells growing for 17 days on borosilicate cover glasses highly up-regulate the expression of bone-specific and of other osteoblast-relevant marker proteins, suggesting that they began to differentiate into mature calcified matrix secreting osteoblasts on this boron-containing substrate. On none of the boron-free substrates (soda lime silicate glass, tissue culture plastic, melt-derived 45S5 or sol-gel derived bioglasses), a similarly significant upregulation was observed when cells were grown for comparable time periods (Figure 3).

Boron has been found previously to stimulate the differentiation of osteoblast precursor and mesenchymal stem cells when added to the culture medium[19, 20], and to enhance bone formation when added to 45S bioglass[21]. This osteogenic effect has been attributed to borate ions in solution that either were added to the culture medium, or leached from the dissolving bioglass after implantation into rat tibias[19,21]. Boron concentrations used ranged from 2% (by weight added as boron oxide to 45S bioglass)[20], to as low as 0.1 ng/ml (added as boric acid to the culture medium)[19]. Since borosilicate cover glasses are not expected to significantly dissolve in cell culture medium, and in addition, we exchanged one-half of the culture medium every other day, in contrast, in our experiments, a much lower concentration of soluble boron is expected, suggesting that either much lower concentrations of borate ions in solution (well below ppm concentrations) are sufficient to stimulate bone cell differentiation; or more likely that bone-precursor cells are able to 'sense' and react to boron that is present in the substrate. Cells clearly can detect differences in substrate composition, as for example indicated by the variable adhesion of MC3T3-E1 cells to TC plastic and borosilicate cover glasses described above (Figure 1).

Silicon, another abundant component of borosilicate, soda lime, melt- and sol-gel derived bioglasses is also discussed as a trace element[16], and hence, MC3T3-E1-cell differentiation observed in our study could potentially also be attributed to the effect of this element. However, although soda lime glasses do contain silicon, no significant differentiation of MC3T3-E1 cells was observed on this substrate; nor on a substrate that does not contain boron or silicon (TC plastic) (Figure 3C).

CONCLUSION

In summary, our results shed new light on the potential role of boron on bone cell differentiation, and suggest a beneficial effect of boron when added to bioglass bone-replacement scaffolds. Future experiments should include a careful analysis of the dissolution of boron from borosilicate and boron-containing bioglasses, and of mechanistic aspects that may allow cells to sense boron in solution, as well as in growth-supportive substrates.

ACKNOWLEDGMENTS

Separate parts of this work were supported by the National Institute of Health (NIH-NIGMS grant R01 GM55725), and National Science Foundation (via Materials World Network (DMR-0602975) and International Materials Institute for New Functionality in Glass (DMR-0844014) programs). The authors thank Dr. Alfred C. Miller for his help with X-ray photoelectron spectroscopy.

REFERENCES

[1] W.C. Billotte, in Biomaterials, J.Y. Wong, J.D. Bronzino, eds., (2006) Chapter 2, CRC Press.

[2] L.L. Hench, 'The story of Bioglass®', *J. Mater. Sci.: Mater. Med.* (2006) 17:967-978.

[3] H.M.M. Moawad and H. Jain, Creation of nano–macro-interconnected porosity in a bioactive glass–ceramic by the melt-quench-heat-etch method., *J. Am. Ceram. Soc* .**90,** 1934-1936 (2007).

[4] A.C. Marques, H. Jain, and R.M. Almeida, Sol-gel derived nano/macroporous monolithic scaffolds, *Euro. J. Glass Sci. Techno.***48,** 65-68 (2007).

[5] A.C. Marques, R.M. Almeida, A. Thiema, M.M. Falk and H. Jain, Sol-gel derived glass scaffold with high pore interconnectivity and enhanced bioactivity, *J. Mater. Res.* **24**, 3495-3502 (2009)..

[6] Y. Vueva, A. Gama, R. Almeida, S. Wang, H. Jain, and M.M Falk, Monolithic glass scaffolds with dual porosity prepared by polymer-induced phase separation and sol-gel, *J. Am. Ceram. Soc.*, **93**, 1945-1949 (2010).

[7] S. Wang, M.M. Falk, A. Rashad, M.M Saad, A.M. Marques, R.M. Almeida, M.K. Marei, H. Jain, Evaluation of 3D nano-macro porous bioactive glass scaffold for hard tissue engineering, *J. Mater. Sci.: Mater. Med.*, in press (2011).

[8] I.D. Xynos , A.J. Edgar, L.D. Butter, L.L. Hench, and J.M. Polak, Ionic products of bioactive glass dissolution increase proliferation of human osteoblasts and induce insulin-like growth factor II mRNA expression and protein synthesis, *Biochem. Biophys. Res. Commun.*, **276**,461-465 (2000).

[9] D.G. Blevins, and K.M. Lukaszewski, Functions of boron in plant nutrition, *Ann. Rev. Plant Physiol. Plant Mol. Biol.*, **49**, 481-500 (1998).

[10] M.T. Gallardo-Williams, R.R Maronpot, C.H. Turner, C.S. Johnson, M.W. Harris, M.J. Jayo, et al., Effects of boric acid supplementation on bone histomorphometry, metabolism, and biomechanical properties in aged female F-344 rats, *Biol. Trace Elem. Res.*, **93**, 155-169 (2003).

[11] M.R. Naghii, G. Torkaman, and M. Mofid, Effects of B and calcium supplementation on mechanical properties of bone in rats, *Biofactors.*, **28**, 195-201 (2006).

[12] A.A. Gorustovich, T. Steimetz, F.H. Nielsen, and M.B. Guglielmotti, A histomorphometric study of alveolar bone healing in rats fed a boron-deficient diet, *Anat. Rec.*, **291**, 441-447 (2008).

[13] F.H. Nielsen, Is boron nutritionally relevant?, *Nutr. Rev.*, **66**, 183-191 (2008).

[14] F.H. Nielsen, C.D. Hunt, L.M. Mullen, and J.R. Hunt, Effect of dietary boron on mineral, estrogen, and testosterone metabolism in postmenopausal women, *FASEB J.*, **1**, 394-397 (1987).

[15] C.D. Hunt, The biochemical effects of physiologic amounts of dietary boron in animal nutrition models, *Environ. Health Perspect.*, **102 (Suppl.7)**, 35-43 (1994).

[16] *Food and Nutrition Board, Standing Committee on the Scientific Evaluation of Dietary Reference Intake, Institute of Medicine, Dietary reference intakes for vitamin A, vitamin K, arsenic, boron, chromium, copper, iodine, iron, manganese, molybdenum, nickel, silicon, vanadium, and zinc.* National Academy Press, Washington, DC (2002).

[17] W.C.A. Vrouwenvelder, C.G. Groot, and K. de Groot, Better histology and biochemistry for osteoblasts cultured on titanium-doped bioactive glass: bioglass 45S5 compared with iron-, titanium-, fluorine-, and boron-containing bioactive glasses, *Biomater.* **15**, 97-106 (1994).

[18] L.L. Hench, Bioceramics: from concept to clinic, *J. Am. Ceram. Soc.*, **74**, 1487-1510 (1991).

[19] S.S. Hakki, B.S. Bozkurt, and E.E. Hakki, Boron regulates mineralized tissue-associated proteins in osteoblasts (MC3T3-E1), *J. Trace Elem. Med. Biol.*, **24**, 243-250 (2010).

[20] N.W. Marion, W. Liang, G.C. Reilly, D.E. Day, M.N. Rahaman, and J.J. Mao, Borate glass supports the *in vitro* osteogenic differentiation of human mesenchymal stem cells, *Mech. Adv. Mater. Struct.*, **12**, 239-246 (2005).

[21] A.A. Gorustovich, J.M. Porto-Lopez, M.B. Guglielmotti, and R.L. Cabrini, Biological performance of boron-modified bioactive glass particles implanted in rat tibia bone marrow, *Biomed. Mater.*, **1**, 100-105 (2006).

[22] R.F. Brown, M.N. Rahaman, A.B. Dwilewicz, W. Huang, D.E. Day, Y. Li, et al., Effect of borate glass composition on its conversion to hydroxyapatite and on the proliferation of MC3T3-E1 cells, *J. Biomed. Mater. Res. A.*, **88**, 392-400 (2009).

[23] A. Yamaguchi, T. Komori, and T. Suda, Regulation of osteoblast differentiation mediated by bone morphogenic proteins, hedgehogs, and Cbfa1, *Endocrine Rev.*, **21**, 393-411 (2000).

[24] G.R. Beck, Jr., B. Zerler, and E. Moran, Phosphate is a specific signal for induction of osteopontin gene expression, *Proc. Natl. Acad. Sci.*, **97**, 8352-8357 (2000).

[25] R. Jain, S. Wang, H. Moawad, M.M. Falk, and H. Jain, Glass Bone Implants: The effect of surface topology on attachment and proliferation of osteoblast cells on 45s bioactive glass, in Engineering

Biomaterials for Regenerative Medicine, edited by S. Bhatia, S. Bryant, J.A. Burdick, J.M. Karp, and K. Walline, *Mater. Res. Soc. Symp. Proc.*, **1235,** RR03-47 (2010).

[26] D. Wang, K. Christensen, K. Chanwla, G. Xiao, P.H. Krebsbach, and R.T. Franceschi, Isolation and characterization of M3T3-E1 preosteoblast subclones with distinct in vitro and in vivo differentiation/mineralization potential, *J. Bone Mineral Res.*, **14**, 893-903 (1999).

[27] L.D. Quarles, D.A. Yohay, L.W. Lever, R. Caton, and R.J. Wenstrup, Distinct proliferative and differentiated stages of murine MC3T3-E1 cells in culture: an in vitro model of osteoblast development, J. *Bone Mineral Res.*, **7**, 683-691 (1992).

[28] C.Y. Chung, A. Iida-Klein, L.E. Wyatt, G.H. Rudkin, K. Ishida, D.T. Yamagushi, and T.A. Miller, Serial passage of MC3T3-E1 cells alters osteoblastic function and responsiveness to transforming growth factor-β1 and bone morphogenic protein-2, *Biochem. Biophys. Res. Comm.*, **265**, 246-251 (1999).

[29] B.L.T. Vaes, K.J. Deschering, A. Feijen, J.M.A. Hendriks, C. Levevre, C.L. Mummery, W. Olijve, E.J.J. van Zoelen, and W.T. Steegenga, Comprehensive microarray analysis of bone morphogenic protein 2-induced osteoblast differentiation resulting in the identification of novel markers for bone development, *J. Bone Mineral Res.*, **17**, 2106-2118 (2002).

[30] T. Saito, H. Hayashi, T. Kameyama, M. Hishida, K. Nagei, K. Teraoka, and K. Kato, Suppressed proliferation of mouse osteoblast-like cells by a rough-surfaced substrate leads to low differentiation and mineralization, *Mat. Sci. Eng.*, **C30**, 1-7 (2010).

Porous Ceramics

HIGH POROSITY IN SITU CATALYZED CARBON HONEYCOMBS FOR MERCURY CAPTURE IN COAL FIRED POWER PLANTS

Xinyuan Liu, Millicent K. Ruffin, Benedict Y. Johnson and Millicent O. Owusu
Corning Incorporated, Corning, NY 14831

ABSTRACT

Corning has developed a novel high porosity honeycomb sorbent that is effective at removing mercury from the flue gas of coal fired power plants. Highly porous activated carbons with pore size distributions in the micro – macro range can be developed by control of the concentration, flow rate and temperature of CO_2 gas. We demonstrate the creation of various porosity profiles and examine the impact of porosity on mercury capture in PRB flue gas. Another factor known to impact mercury capture on carbon surfaces is the presence of sulfur species. Studies are presented that assess the effectiveness of thiols, thiophenes, elemental sulfur and sulfides on mercury capture efficiency. Lastly, carbon surface modifications to reduce the accumulation of sulfur dioxide on the honeycomb during exposure to mercury containing simulated PRB flue gas is shown to be effective in the reduction of sulfate accumulation.

INTRODUCTION

Mercury is one of the nation's highest priority pollutants, such that its emission has been of great concern due to its high toxicity, tendency to bioaccumulate and difficulty to control [1]. Coal-fired power plants are the major source of mercury emission [2]. In the US alone, coal-fired power plants emit a total of about fifty metric tons of mercury into the atmosphere annually, and there is approximately 1470 tons/year of mercury released worldwide. Mercury is present in the flue gas of coal-fired power plant in several forms: elemental (Hg^0), oxidized ($HgCl_2$ and HgO), and particulate bound mercury [3]. Once released into the environmental atmosphere, mercury can transport and cycle among the air, water and land, then it can persist in the ecosystem and eventually enter and accumulate in the food chain. It can be converted by microorganisms into an organic form, methyl mercury, which is a neurotoxin that bioaccumulate in fish, animals and mammals [4]. The main exposure route for humans is the consumption of fish that are contaminated with mercury. Studies indicate that exposure to mercury can have adverse effects on kidneys, the central nervous system, and it may induce severe gastrointestinal and respiratory damage [5].

Several technologies have been developed to control mercury emission including electrostatic precipitators (ESPs), fabric filters, flue gas desulfurization (FGD), spray dryer adsorption (SDA), and activated carbon injection (ACI) [6-8]. Among these methods, activated carbon injection is commonly considered to be the simplest and most promising technology for gaseous mercury removal from coal-fired power plants that do not have wet scrubbers (about seventy-five percent of all plants) [1, 8]. However, its current application to mercury capture is very limited due to low sorbent capacity at elevated temperatures, flue gas interactions with activated carbon, relatively high cost for sorbent materials, and problems with fly ash disposal [1, 9, 10].

It has been shown that a packed or fixed activated carbon bed can remove mercury with an efficiency as high as 95% with the additional benefit of preserving lucrative fly ash sales. But fixed bed systems are generally an interruptive technology requiring frequent sorbent bed replacement and it is usually associated with a high pressure drop which significantly reduces energy efficiency [11, 12]. Previous studies have reported that, compared with untreated carbon, introduction of sulfur onto carbon surfaces can produce more stable products and improve mercury capture efficiency significantly [7, 13]. It was also found that the mercury uptake capacity was determined by sulfur content, sulfur forms, sulfur distribution, surface area and pore structure of the carbon sorbents [13-15].

Corning has a history of innovation that includes a 30+ year old business of honeycomb substrates for automotive applications. In this paper, we present an extension of that technology to carbon platforms, and demonstrate the advantages and challenges of high porosity activated carbon honeycombs for use as mercury capture devices. Carbon honeycombs are shown to be an effective fixed bed technology for application in coal fired power plants. We demonstrate the application of the porous carbon honeycombs for mercury capture focusing on the impact of porosity, sulfur species, and acid gases. We also demonstrate carbon surface chemistry modifications to remediate acid gas accumulation.

EXPERIMENTAL

Porosity

Various CO_2 flow rates and activation temperatures were used to achieve porosity in the micro, meso and macro ranges. The activation time required for each sample was dependent on flow rate and temperature conditions. In general, CO2 flow rates ranged from 0.5 – 6 SCFH and temperatures were in the range of 650°C - 900°C.

Sulfur species

Activated carbon substrates were first dried in 130°C oven for 20 minutes and sonicated in 15% H_2SO_4 for 10s, and then dried by air knife and then in an 150°C oven overnight. Activated carbon honeycombs were impregnated with different species of sulfur: elemental sulfur, metal sulfides, thiol and thiolpehene. Aryl sulfonate groups were introduced onto the surface of a carbon substrate (S1 or S2) modified from a patent reported method [16]. Each carbon substrate was soaked in sulfanilic acid solution, and then sodium nitrite was added to the solution. The reaction was kept in a water bath at 70°C for 2hr. The treated carbon substrate was removed, washed thoroughly with DI H_2O to remove excess acid and salts, dried in an oven at 110°C overnight, and then dried at 180°C for 30-60 minutes.

SO2 remediation

The SO_2 adsorption of treated and untreated samples was measured in a simulated flue gas mixed through a gas delivery system, nominally providing 10% CO_2, 7.5% O_2, 2% H_2O, 250 ppm SO_2, 300 ppm NO, 10 ppm NO_2, with the balance N_2. FTIR measurements were referenced against a nitrogen background taken immediately prior to sample measurement. The concentrations of water soluble sulfate in the samples were measured by ion chromatography using a Dionex ICS3000 system equipped with a Dionex conductivity detector.

Mercury capture measurements

The carbon honeycombs were measured for mercury uptake and mercury breakthrough under dynamic conditions. Testing was conducted with 3-23 days of exposure. Inlet elemental Hg concentration was 10-25 g/m^3. Simulated Powder River Basin flue gas was composed of 6% O_2, 12% CO_2, 7~10% H_2O, 400ppm NO, 3ppm HCl, 400ppm SO_2.

Field Test measurements

Samples were exposed to real Powder River Basin (PRB) flue gas at 15000 hr^{-1} space velocity. The inlet and outlet Hg concentration (elemental and total) were monitored using a Tekran mercury analyzer to assess the Hg capture efficiency and usable life of the parts. Samples were exposed to flue gas for set periods of time. Following exposure, samples were analyzed for the Hg present in the part as well as the chemical makeup of the carbon substrate after exposure to flue gas.

RESULTS AND DISCUSSION

Highly porous activated carbons with pore size distributions in the micro – macro range are key for optimizing mercury adsorption potential. Basically for carbonaceous systems, the porous structures are generated by carbonization and activation. There are three process levers for controlling porosity: CO_2 concentration, temperature and flow rate. By varying each of these three conditions, porosity in micro, meso and macro regions can be generated with the activation times range from 2hrs to 48hrs. As shown in Figure 1, three extremes of pore structure created: meso-micro (2 nm – 50 nm), macro-micro (2 nm and >50 nm) and macro-meso-micro (broad range).

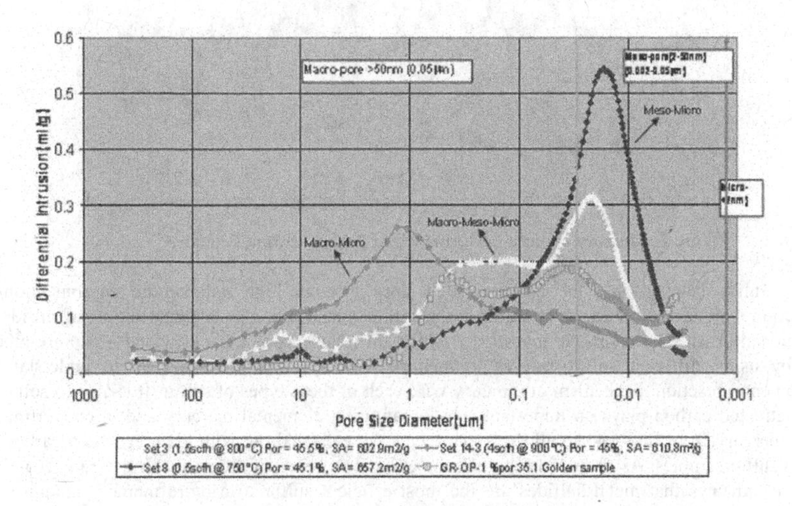

Figure 1 Porosity generated in activated carbon honeycombs

Those porous carbon substrates were tested in PRB flue gas, mercury concentration at the outlet was measured either during the day for 4 hours or overnight. As demonstrated in Figure 2, meso-micro pore structure (square) performed better days 1-9, however after 11 days of testing all 3 samples had same capture efficiency (70%), suggesting that porosity is not a primary factor on performance.

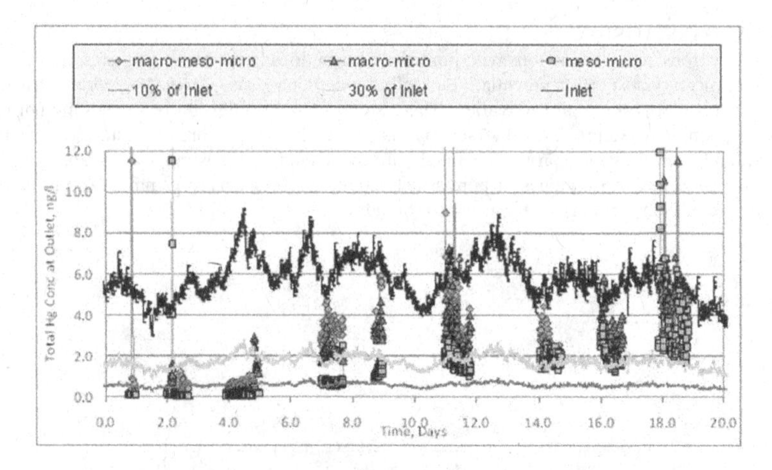

Figure 2. Mercury capture performance of porous carbon structure

Sulfur batched into an extrusion that goes through high temperature carbonization and activation may have three forms of sulfur in the carbon substrates: elemental sulfur, metal sulfide, and organic sulfur. In this report, we investigate which sulfur species is most efficient to capture mercury and by using different sulfur species to capture mercury, it will be helpful to understand the fundamental reaction mechanism of mercury with each of these types of sulfur. It is also hypothesized that activated carbon plays an important role in capturing elemental mercury and in converting it to ionic mercury from flue gas, so all the experiments were carried out with plain activated carbon (S1) for all sulfur samples. As shown in Figure 3, the efficiency trends of different sulfur forms for mercury capture indicates that metal sulfides are the most efficient sulfur to capture mercury among all the tested sulfur species, and the mercury capture efficiency order is demonstrated as: metal sulfide >> elemental sulfur~ thiol > thiophene. The S1 only, S1+thiol and S1+S samples share the same performance trajectory. The S system showed signs of recovering some efficiency but the other two were stopped, maybe before they could recover. Note that the recovery of the mercury capture efficiency for the sulfur containing sample is very interesting, and longer term testing would be needed to understand all aspects of its mercury storage capacity.

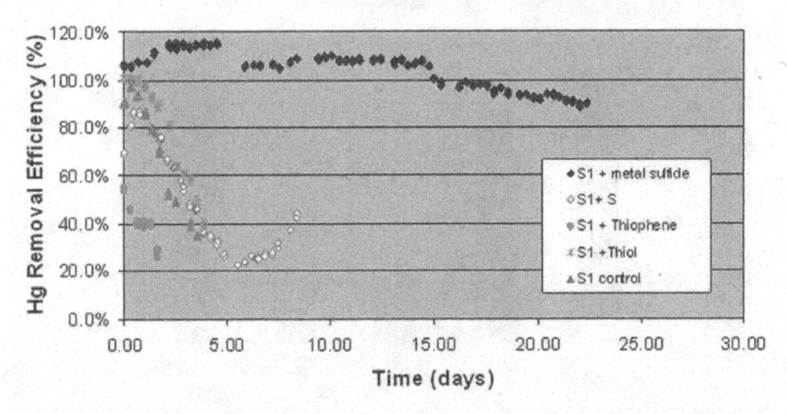

Figure 3 Mercury capture efficiency comparison for all the sulfur forms

Figure 4 presents the anion accumulation results from ion chromatography measurements of different sulfur samples exposed to Hg-laden simulated flue gas for various days. The exposure time for S1+ metal sulfide was ~3 weeks, ~1 week for S1+S sample, and it was 3~4 days for S1+ thiophene, S1+thiol and S1 control samples. Sulfate poisoning may contribute to the declining mercury capture efficiency of S1+ metal sulfide after two weeks exposure (Fig. 3). Compared with S1 control, after similar exposure time S1+thiophene sample shows a relatively low sulfate accumulation, and this might be due to the carbon surface active sites blockage by insoluble thiophene molecules. These IC data suggest that high surface area of active carbon contributed significantly to SO_2 adsorption and sulfate accumulation, which may lead to early mercury capture breakthrough and short sample life time. Therefore, to achieve desirable long term high efficiency mercury capture, treatments are needed to effectively decrease sulfate formation or adsorption on carbon substrates.

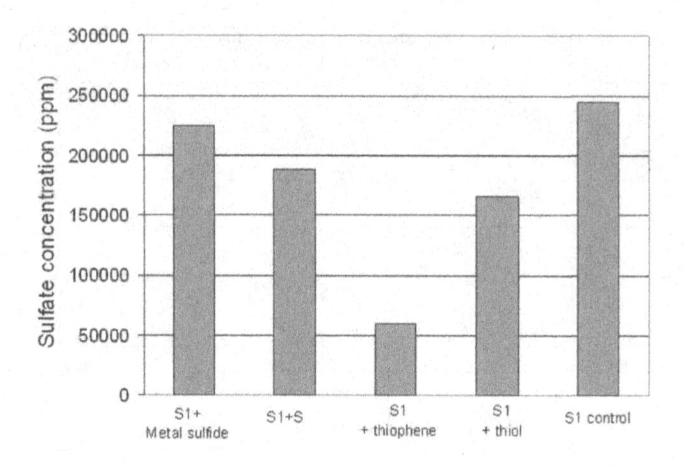

Figure 4. Sulfate concentration after exposure to Hg-laden simulated flue gas for various days

Then we investigated how sulfuric acid can affect mercury capture performance. Samples were sonicated in 15% H_2SO_4 for 10 seconds. Figure 5 presents the mercury removal efficiency of control and sulfuric acid treated samples. The control sample has 8 days to breakthrough, however, it only took 3 days for the sulfuric acid treated sample to breakthrough, indicating that sulfuric acid shortens lifetime of capture efficiency.

Coal-fired combustion flue gases contain several acid gases including SO_2 and SO_3, NO and NO_2, and HCl. These acid gases have been shown to affect the performance of activated carbon sorbents that have worked well under other conditions. Olson et al reported that when SO_2 was added to gas mixture with NO_2, mercury capture breakthrough occurred after 1hr at 225 F with a relatively steep curve, and it increased to 100% or greater emission ca 2hr, indicating no more mercury adsorption and release of mercury adsorbed earlier [17]. We also observed similar adsorption inhibition and even desorption of mercury from carbon when SO_2 was added into the flue gas as shown in Figure 6, which was attributed to the competition for the same adsorption sites on carbon [3, 18]. Thus, the instability of activated carbon sorbents in an acidic flue gas environment adversely limits the capacity and the utility of these sorbents.

The reaction of SO_2 with carbon in the presence of O_2 and H_2O at relatively low temperatures (20-150 °C) involves a series of reactions that leads to the formation of sulfuric acid as the final product. The overall reaction is $SO_2 + 1/2O_2 + H_2O + C \rightarrow C\text{-}H_2SO4$. In the literature the following reaction sequence has usually been presented [19-22].

$$C + SO_2 \rightarrow C\text{-}SO_2$$
$$C + 1/2O_2 \rightarrow C\text{-}O$$
$$C + H_2O \rightarrow C\text{-}H_2O$$
$$C\text{-}SO_2 + C\text{-}O + C\text{-}H_2O \rightarrow C\text{-}H_2SO_4$$

It implies that SO_2, O_2, and H_2O are all adsorbed on the surface of the carbon in close enough proximity and in the proper steric configuration to react and form H_2SO_4. Experimental parameters that may affect SO_2 removal by carbon include flue gas temperature, concentration of other pollutants in the flue gas, flow rate, amount of O_2 and H_2O in the flue gas, and the physical and chemical properties of the carbon such as surface area and functional groups.

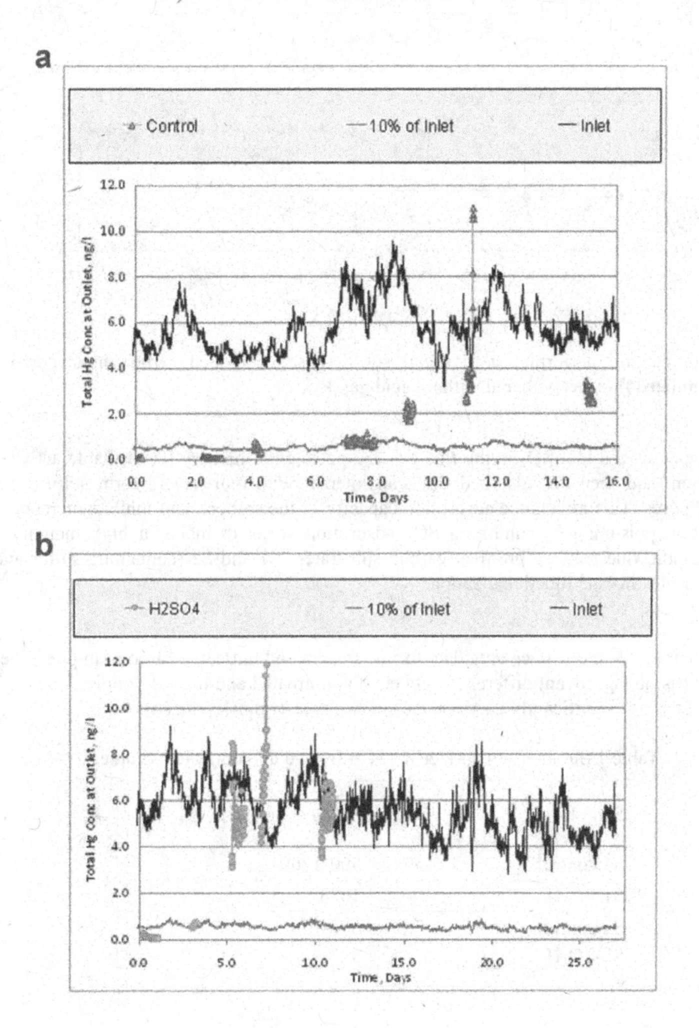

Figure 5 Mercury capture efficiency comparison for control and sulfuric treated samples

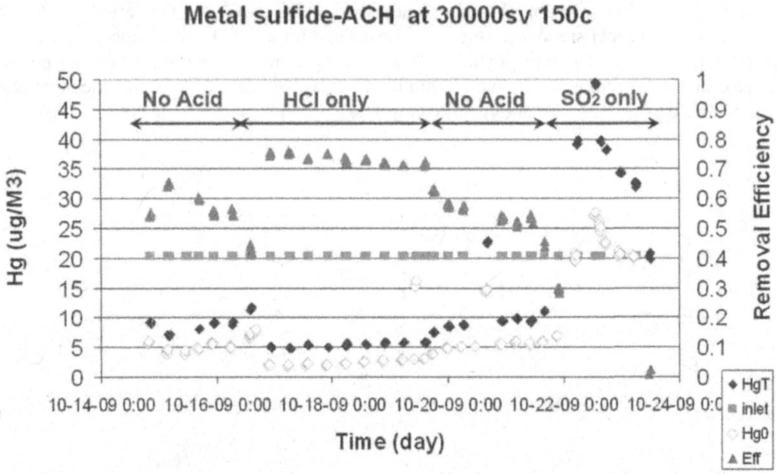

Metal sulfide-ACH at 30000sv 150c

Figure 6 Hg capture performance of metal sulfide-based activated carbon honeycomb sample exposed to simulated flue gas with and without acid gases

It is hypothesized that SO_2 in the flue gas competes with mercury for available active sites on activated carbon, and then the adsorbed SO_2 can interact with moisture to form sulfuric acid that penetrates the pores. This affects the adsorption capacity of the sorbent, and inhibits mercury capture. To avoid carbon poisoning by inhibiting SO_2 adsorption so as to maintain high mercury capture efficiency in acidic flue gas, we pre-treat carbon substrates (S1 and S2 (containing sulfur and metal sulfide)) with aryl sulfonate functionalization.

Table 1 presents the surface area data for the treated and untreated carbon samples. The results show that there is no significant difference between the untreated and treated samples, suggesting that this treatment does not significantly decrease the surface area or block pore structures.

Table 1 Summary of surface areas of treated and untreated samples

Samples	Surface Area (m^2/g)
S1 control	900-1200
S1-SO_3H	1085
S2 control	600-750
S2-SO_3H	593

Figure 7 shows the SO_2 adsorption of the treated and untreated samples by FTIR spectroscopy. For the S1 control, the SO_2 concentration dropped dramatically from ~233 ppm to 110 ppm whereas it only decreased from 240 ppm to 190 ppm for S1-SO_3H. In addition, it took relatively shorter time for SO_2 to recover to its original concentration for the sulfonated sample compared with the untreated S1. This effect was more significant for the S2 samples. For the treated S2 sample, the SO_2 concentration only decreased from 242 ppm to 216 ppm and it recovered to the original concentration in 5 mins. For the S2 control sample other hand, the SO_2 concentration reduced from 245 ppm to 162 ppm and it took much longer time to recover back. It is very clear that the sulfonation treatment can effectively inhibit adsorption of SO_2 in the flue gas.

As mentioned in the introduction section, coal-fired combustion flue gases contain several acid gases including SO_2, SO_3, NO, NO_2, and HCl. These acid gases have been shown to degrade the performance of activated carbon sorbents that have worked well under other conditions. Some components can compete with the mercury for available adsorption sites. In other cases, mercury reactive sites in the sorbents can be deactivated. The objective of this short-term, accelerated Hg uptake test was to evaluate the impact of the surface modification treatment on Hg capture efficiency. It is desirable that the treatment will prevent or minimize competition with Hg for available adsorption sites which will then allow the material to perform according to its inherent capability for capturing Hg. For this treatment to be effective, several factors are important, namely, surface chemistry, pore structure, flue gas residence time and temperature.

The treated and untreated S2 samples were exposed to simulated flue gas with 10-25 $\mu g/m^3$ Hg (0) as inlet in house for 7 days with a high space velocity of 30,000/hr, and the outlets of total mercury and elemental mercury concentrations were measured at 150°C respectively. The treated and untreated GKW samples were exposed for only 3-4 days due to their early breakthrough.

Figure 8 presents the mercury removal efficiency of treated and untreated samples. As seen in Fig.8a, the inlet mercury concentration is ~10 $\mu g/m^3$, and the background mercury concentration is almost 0 $\mu g/m^3$, both of the inlet and background data were measured before and after sample exposure. It can be seen that when sulfonated S2 was introduced into the system, the mercury concentration decreased significantly to 0 ppm, where the diamond represents the total mercury outlet and the elemental mercury outlet is shown in hollow diamond, respectively. The efficiency of mercury capture was then calculated based on the inlet, outlet and background mercury concentrations, and it was then plotted as triangle in this figure. After 7 days exposure to simulated flue gas, the treated S2 sample still maintained capture efficiency close to 90%. The capture efficiency of the S2 control sample is presented in Figure 8b. It can be seen by comparing Figures 8a and 8b that the sulfonated sample has comparable performance to the untreated sample indicating that the treatment did not interfere with the Hg adsorption and it does not affect the initial mercury capture efficiency.

Figure 9 demonstrates the mercury capture performance of sulfonated S1. It is seen that the efficiency started declining only after 1 day exposure and it decreased to as low as 20% in 3 days. It is also noted that the elemental mercury concentration was very low during the whole exposure period, suggesting an effective oxidation of Hg (0) by activated carbon, though the total mercury concentration increased significantly over time. One hypothesis of breakthrough is that due to absence of sulfur or sulfide in the carbon substrates, which can lead to Hg (II) release after adsorption and oxidization of Hg (0). To test this hypothesis, sulfur or metal sulfide, which can capture the released Hg (II), may be placed in series after the treated S1 sample in the mercury performance test to see if it has longer breakthrough time. Another possible hypothesis is that the sulfonation groups are not enough to show significant difference between treated and untreated samples, though it inhibited SO_2 adsorption according to FTIR data. We may apply longer soaking time and reaction time to obtain higher density of functional groups on the carbon substrate surfaces. While the real impact of the surface modification treatment on longevity of the materials can only be assessed after a longer period of exposure to simulated flue gas or accelerated test in real flue gas, our preliminary findings at least indicate that this treatment can increase the effectiveness and the life of the activated carbon sorbent materials for adsorption of mercury by reducing adsorption of SO_2.

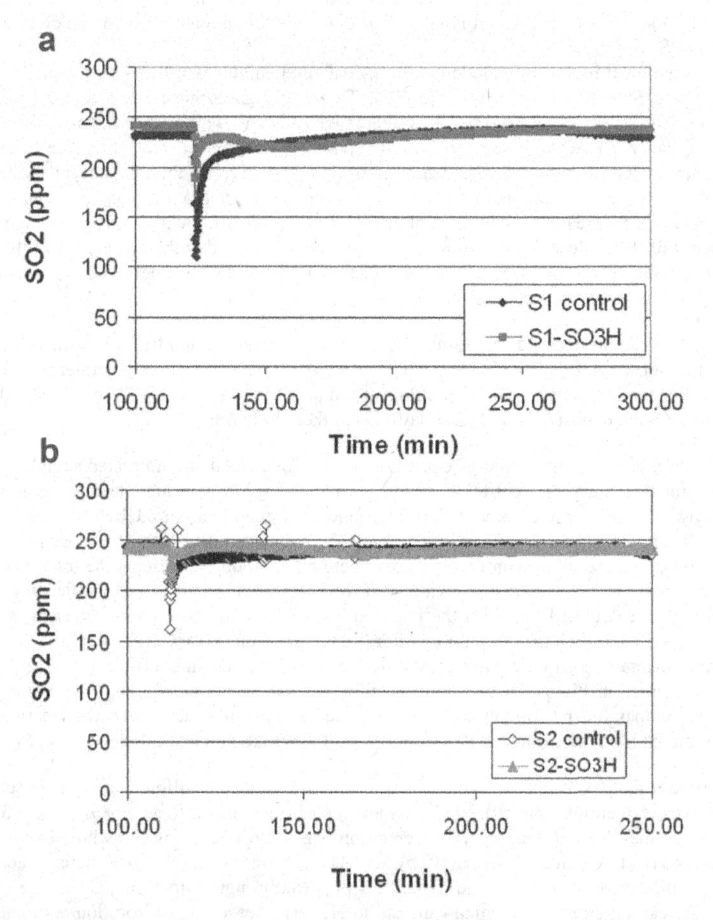

Figure 7 FTIR spectrum illustrating SO_2 adsorption on treated and untreated GKW (left) and GIL (right) samples

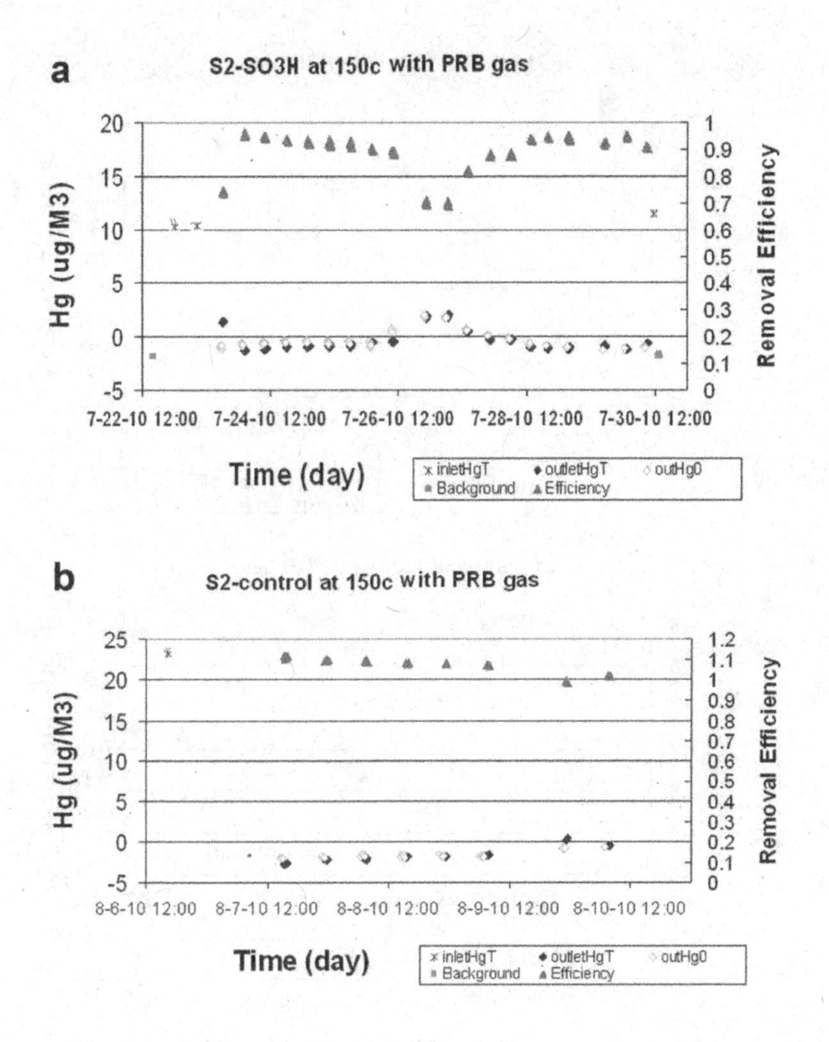

Figure 8. Hg removal efficiency of sulfonated S2 and untreated S2 activated carbon samples.

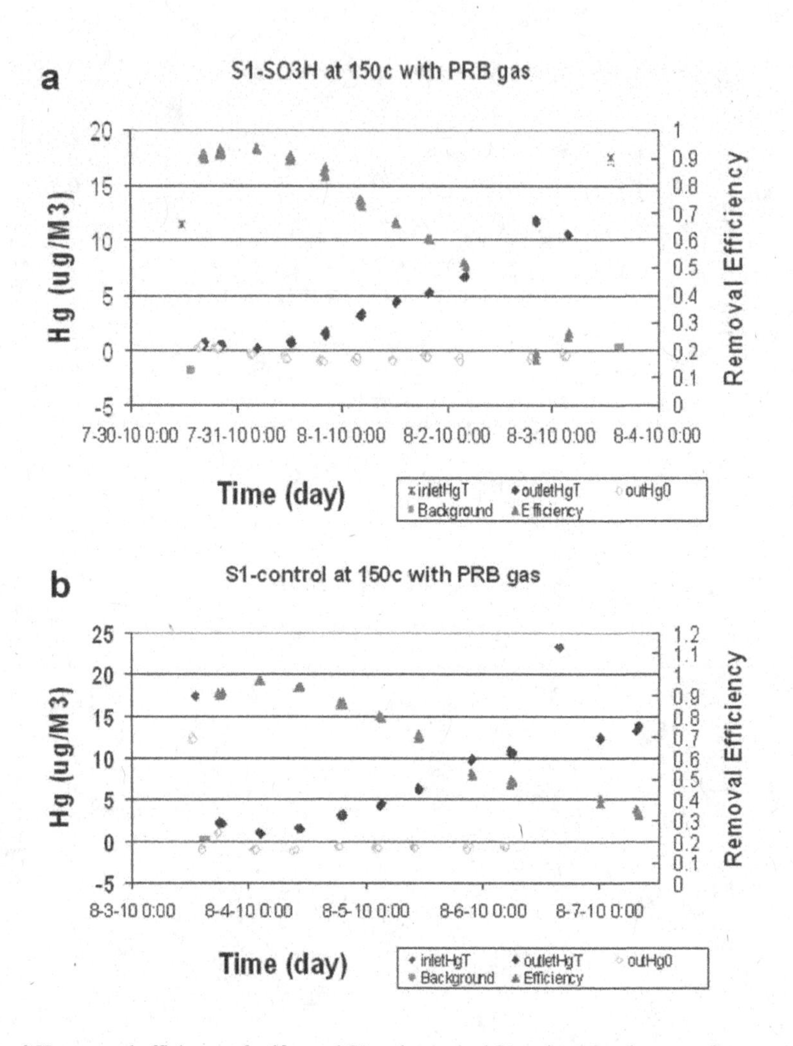

Figure 9 Hg removal efficiency of sulfonated S1 and untreated S1 activated carbon sample.

Figure 10 is a plot of ion chromatography (IC) data showing sulfate contents of treated and untreated samples exposed to Hg-laden simulated flue gas for various days. It is worth mentioning that for direct comparison, both the treated and untreated samples were subjected to the same exposure conditions (i.e exposure time and space velocity) It is evident from Figure 10 that there is much less sulfate accumulation on the treated samples than untreated ones for both S1 and S2 carbon substrates.

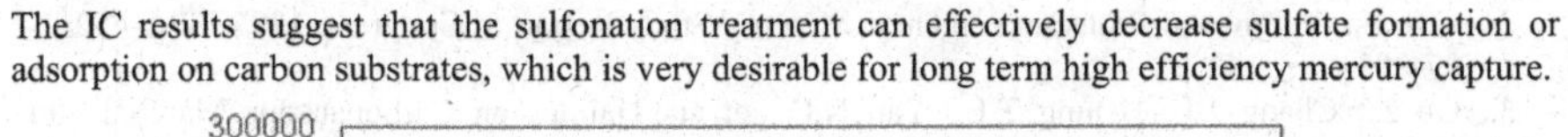

The IC results suggest that the sulfonation treatment can effectively decrease sulfate formation or adsorption on carbon substrates, which is very desirable for long term high efficiency mercury capture.

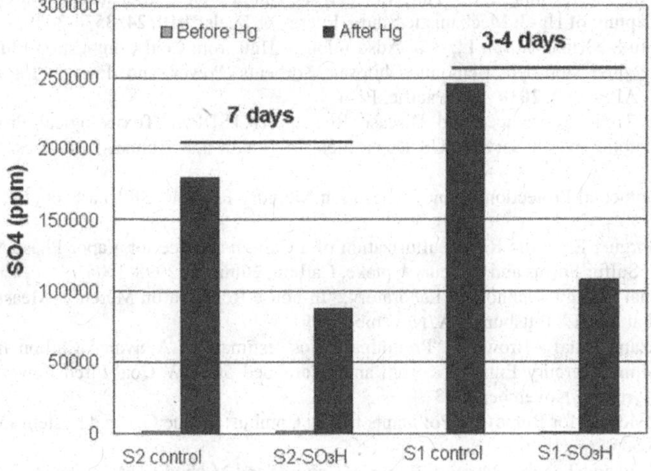

Figure 10. Mercury concentration on untreated and treated samples after exposure to simulated flue gas.

CONCLUSION

Here in this paper, we demonstrated that porous structured carbon can be generated by activation, but porosity is not a primary factor on performance. Study of different sulfur species in activated carbon indicates that metal sulfides are most efficient at mercury capture, buildup of sulfuric acid decomposes the metal sulfide and causes early breakthrough. It is found that presence of SO_2 in flue gas shortens the lifetime of activated carbon substrates. Sulfonation on carbon surface can significantly reduce SO_2 adsorption and sulfate accumulation without affecting its mercury removal efficiency.

ACKNOWLEDGEMENTS

Technical support and discussion of Elizabeth Vileno, Steven Dawes, Ezra Yarnell, Li Liu, Joseph Ward, Ifeoma Nwakalor, David Stalbird, Lisa Hogue, Todd Fleming, Wanda Walczak, Scott Kabel, Robin Balcom, Pete Robbins, Jason Schoonmaker and Kishor Gadkaree from Corning Incorporated are gratefully acknowledged.

REFERENCES

1. Suriyawong A., Smallwood M., Li Y., Zhuang Y. and Biswas P. Mercury Capture by Nano-structured Titanium Dioxide Sorbent during Coal Combustion: Lab-scale to Pilot-scale Studies. Aerosol and Air Quality Research, **2009**, 9, 394-403.

2. U.S. Environmental Protection Agency, Mercury Study Report to Congress, **1997**, EPA-452/R-97-003.
3. Qu Z., Chang J.J., Hsiung T.L., Yan N.Q., et al Halides on Carbonaceous Materials for Enhanced Capture of Hg-0: Mechanistic Study. Energy & Fuels **2010**, 24, 3534-3539.
4. Zhang A., Hu S., Xiang J. Sun L,, et al Adsorption of Hg0 from Coal Combustion Flue Gases by Novel Iodine-Modified Bentonite/Chitosan Sorbents Power and Energy Engineering Conference (APPEEC), **2010** Asia-Pacific, P1-4
5. Agency for Toxic Substances and Disease Registry (ATSDR), "Toxicological Profile for Mercury", Public Health Service, US Department of Health and Human Resources, Atlanta, GA **1999**
6. U.S. Environmental Protection Agency, Clean Air Mercury Rule, 40 CFR Parts 60, 63, 72, and 75, **2005**.
7. Feng W., Borguet E., Vidic R.D., Sulfurization of a Carbon Surface for Vapor Phase Mercury Removal-II: Sulfur Forms and Mercury Uptake. Carbon, **2006**, 44, 2998-3004.
8. DOE-National Energy Technology Laboratory, "In-house Research on Mercury Measurement and Control at NETL" Pittsburg, PA, November **2001.**
9. Hoffmann J.and Ratafia-Brown J. "Preliminary Cost estimate of Activated Carbon Injection for Controlling Mercury Emissions from an Unscrubbed 500MW Coal-Fired Power Plant", DOE- NETL report, November **2003**
10. Chang R., "Method for Removing Pollutants from a Combustor Flue Gas and System for Same, US patent 5,505,766, **1996**.
11. Gadkaree K. P. and Tao T., "Mercury Removal Catalyst and Method of Making and Using Same", US Patent, 6,258,334, **2001**.
12. Shi Y., Gadkaree K. P. and Dawson-Elli D. F., "Activated Carbon Honeycomb for Highly Efficient Hg Removal", Proceedings of the 30th International Technical
13. Liu W., Vidic R. Optimization of High Temperature Sulfur Impregnation on Activated Carbon for Permanent Sequestration of Elemental Mercury Vapors. Environ. Sci. Technol. **2000**, 34(3), 483-488.
14. Liu W., Vidic R.Optimization of Sulfur Impregnation Protocol for Fixed-Bed Application of Activated Carbon-Based Sorbents for Gas-Phase Mercury Removal. Environ. Sci. Technol. **1998**, 32(4), 531-538.
15. Hsi H-C, Rood M.J., Rostam-Abadi M., Chen S. and Chang R. Effects of Sulfur Impregnation Temperature on the Properties and Mercury Adsorption Capacities of Activated Carbon Fibers (ACFs). Environ. Sci. Technol. **2001**, 35(13), 2785-2791
16. Belmont J.A., Johnson J.E. and Adams C.E. Ink Jet Ink Formulations Containing Carbon Black Products, US Patent, 5,571,311, **1996**.
17. Olson E.S., Dunham G.E., Ramesh K. Sharma R.K.. Stanley J. Miller S.J. Mechanisms of Mercury Capture and Breakthrough on Activated Carbon Sorbents. American Chemical Society National Meeting, Washington DC, **2000**, 886-889.
18. Liu S.H., Yan N.Q., Liu Z.R., Qu Z et al Using Bromine Gas To Enhance Mercury Removal from Flue Gas of Coal-Fired Power Plants Environ. Sci. Technol. **2007**, 41, 1405–1412.
19. Richter, E.; Knoblauch, K.; Jungten, H. Mechanisms and Kinetics of SO_2 Adsorption and NO_x Reduction on Active Coke Gas Sep. Purif. **1987**, 1, 35-43.
20. Jungten, H.; Kuhl, H. Chem. Phys. Carbon **1989**, 22, 145.
21. Richter E. Carbon Catalysts for Pollution Control Catal. Today **1990**, 7, 93.
22. Tsuji, K.; Shiraishi, I. In Proceedings of Electric Power Research Institute SO_2 control Symposium; Washington, DC, **1991**, p 307.

NOT ALL MICROCRACKS ARE BORN EQUAL: THERMAL VS. MECHANICAL MICROCRACKING IN POROUS CERAMICS

Giovanni Bruno
Corning Incorporated, S&T, CS&S, SP-FR06
Corning, NY, 14831, USA

Alexander M. Efremov
Corning SNG, M&S, CSC
St.Petersburg, 194021, Russia

Chong An
Corning Incorporated, MTE
Corning, NY, 14831, USA

Seth Nickerson
Corning Incorporated, ET, DV
Corning, NY, 14831, USA

ABSTRACT

Porous microcracked ceramic materials present interesting peculiarities such as very low thermal expansion and very high strain tolerance. Those make them excellent thermal shock resistant materials.

Microcracking is generally generated during cooling from the firing/ceramming temperature, due to a large crystal thermal expansion anisotropy. We have investigated some of the peculiar properties of porous ceramics such as the (partly negative) thermal expansion, the hysteresis of the Young's modulus as a function of temperature, and its non-linearity as a function of applied uniaxial compressive load. Microcracked porous ceramics, such as β-eucryptite, cordierite and aluminum titanate have been taken as examples, and we have compared them with non-microcracked porous ceramics such as silicon carbide and alumina.

We have observed that the Young's modulus and the thermal expansion of microcracked ceramics cycle as a function of temperature, following the same curve; moreover, the Young's modulus has a strong variation (different for each material) as a function of applied load. This, linked to the observation that thermally-induced microcracks have a particular crystallographic orientation, has led us to distinguish two kinds of microcracks: thermal and mechanical. The latter are generated during loading, as extension of existing ones, or fresh, and may not be reversible as the former. Indeed, the time-dependence analysis of the stress-strain curves has allowed us to produce further evidence that the two phenomena coexist in porous microcracked ceramics. Most of the non-linear phenomena do not appear in non-microcracked ceramics.

Finally, the integrity factor modeling work has allowed rationalizing the phenomenon.

Keywords: porous ceramics; neutron diffraction; laser ultrasonics; microcracking; Young's modulus

INTRODUCTION

Porous microcracked ceramic materials present interesting peculiarities with respect to their non-microcracked analogues. Those properties, such as very low thermal expansion and very high strain tolerance, make them excellent candidates for applications where thermal shock resistance is the primary requirement. Typical porous microcracked ceramics are mullite, β-eucryptite, cordierite and aluminum titanate. All of them are currently used for membrane, catalytic substrate and filter applications. The thermal shock resistance is a combination of thermal conductivity and expansion with mechanical properties such as strength and Young's modulus. The variation of these properties with temperature and/or with applied stress as well as with microstructural parameters, such as porosity and microcracking is of utmost relevance.

The latter aspect has received little attention so far, as the Young's modulus of ceramics is generally considered a constant and the stress-strain response fairly linear. However, some inorganic compounds are already known to exhibit a dependence of elastic properties upon applied stresses. The most relevant examples come from the geophysical field. Porous rocks often have shown to possess elastic moduli, which are not constant but increase with increasing stress. Brown et al.[1] stated that the use of classical linear elasticity (i.e. constant modulus) can lead to erroneous predictions of the deformations and of the initiation and extent of failure around underground excavations. Yu and Dakoulas[2] found that elastic properties of a soil depend on the stress state, and developed a model for those dependencies. Heap et al.[3] found a variation of both Young's modulus and Poisson's ratio of basalt rocks as a function of cycle number in alternate loading. The explanation they gave is related to damage and microcracking. Interestingly, they found by acoustic emission, that events can be recorded only if the load in a particular cycle exceeds any of the loads applied in previous cycles. They also found a non-linear stress-strain behavior in each of the stress-strain cycles.

Industrial ceramics generally have a complex microstructure. Even in the case where they are single-phased, the crystal symmetry is very low and the anisotropy of the crystal thermo-elastic properties is high. Consequently, there is often a mismatch between the thermal expansions of different crystal axes, so that microcracks are created upon cooling from the fabrication temperature, almost regardless of the processing route. This influences the elastic properties and the internal stresses locked in the microstructure. The subject has been treated in preceding works[4,5]. There, the role of microcracking has been elucidated and a physical scenario has been proposed to explain how the amount of microcracks can radically change the stress-strain response of a porous ceramic.

In this work, the investigation initiated in[4] is extended, and further experimental evidence of the properties and of the evolution of microcracking in porous ceramic materials is shown. Modeling work is also added to corroborate the experimental observations and the proposed explanation. A comparison with non-microcracked materials is made, and the fundamental differences underlined. We finally conclude showing how thermally and mechanically induced microcracks fundamentally differ.

EXPERIMENTAL TOOLS

A large number of experimental techniques have been used to produce the amount of data shown in the following. Some of them just brought further evidence to well-known data. We will therefore refer to the original work and describe in this paragraph only the techniques where the majority of the results stem from. The laboratory equipment used to generate the data mentioned in this work was the following:

A Netzsch DL402 single pushrod dilatometer for macroscopic thermal expansion measurements;
An Autopore 9520 Micromeritics for porosity measurements
An Instron stress rig with 100 and 250 kN load cells for uniaxial compressive testing
A in-house built device for sonic resonance Young's modulus measurements

The two techniques, which have been more massively used in this work and have less of a conventional character, are described in the following.

Neutron Diffraction

Neutron Diffraction strain analysis has now become a widely accepted tool for internal strain measurements. It is based on the measurement of peak shifts with respect to an unstrained reference. In the case in-situ measurements, the strain reference point is taken at zero applied stress ($\sigma = 0$) and the strain at any applied load is calculated according to .

$$\varepsilon(\sigma)=\frac{d(c)-d(c=0)}{d(\sigma=0)} \qquad (1)$$

In eq.(1) d is the lattice parameter, as calculated by Rietveld refinement[6] at pulsed sources using the time-of-flight technique[7], but can be just one interplanar spacing, if a steady state reactor source is used. A sketch of the measurement set-up is shown in Fig.1, for the case of the SMARTS (Spectrometer for materials at temperature and stress) at the LANSCE, LANL, Los Alamos, NM, USA. Diffraction patterns are acquired while the load is held at several constant values (the experiment is therefore run in load control). The set-up with two opposite detectors allows measuring the axial and the transverse microscopic response (strain). Further details on the technique are available in textbooks[8].

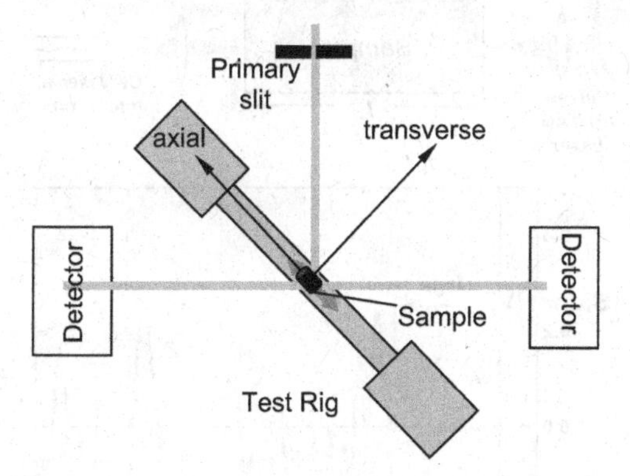

Figure 1. Schematic diagram of the time-of-flight Neutron Diffraction technique for internal strain measurements.

Laser Ultrasonics

The Laser Ultrasonic technique[9] has been developed as a non-contact method for the measurement of mechanical properties of materials. The basic principle is described in the Figure 2. The shock wave is generated by ablating the sample surface with a high energy laser pulse. The ultrasonic pulse at the other side is detected by two-wave mixing technique of continuous-wave laser. A laser ultrasonic system by Intelligent Optical System Inc. is used in this work.

The ultrasonic time-of-flight (ToF) for a known distance of sample yields the speed of sound in the shortest path between the generation and the detection points. It can be used to obtain the elastic modulus E with the mass density ρ and structural factor $f(\nu)$ of a sample, where is the Poisson's ratio, according to the following relation:

$$E = \rho \cdot f(\nu) \cdot \left(\frac{L}{t_{TOF}} \right)^2 \tag{2}$$

where L is the path length, t_{TOF} is the signal time-of-flight (see Fig.2).

In porous ceramic samples with microcracks the Poisson's ratio is assumed to be small and the structural factor $f(\omega)$ becomes 1.

Stress is applied on a sample with a mechanical vise. Since the sample surface is not perfectly flat, alumina fiber felt-like mats are inserted between the sample surface and the vise metal. The stress is calculated over the sample area including pores and open channels (in other words, the applied stress is not corrected for the open frontal area and represents an effective, homogenously smeared stress). For a honeycomb structured sample, the local stress will be much higher than the stress value presented here. In every step of stress-up or -down, the measurement is waited until the stress is completely stabilized without stress-overshooting. A full load/unload cycle measurement takes about 1-2 hours.

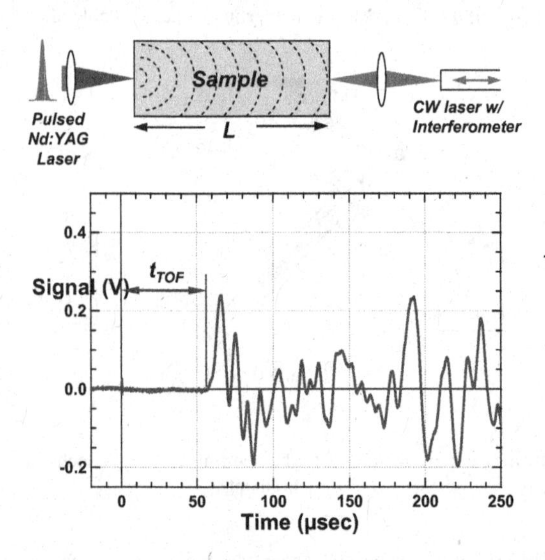

Figure 2. Schematic diagram of Laser Ultrasonic technique for time-of-flight measurements, and an example of time-domain spectrum of Laser Ultrasonic response

MATERIALS

A large spectrum of porous materials has been chosen. Microcracked materials used in this investigations were reaction-sintered β-eucryptite, cordierite and aluminum titanate. They have all been prepared by firing an extruded batch mixture of crystalline raw materials. A general overview of the batch composition and of the firing conditions is given in Table I for the three materials.

All of them were extruded with a cellular structure, similar to that of a diesel particulate filter. In parallel, non-microcracked materials were also investigated: alumina and silicon carbide. The first was sintered from initial powder (1 μm particle size), the second was extruded as cellular material, analogous to the microcracked ceramics quoted above.

Table I. Batch components and firing conditions for the three investigated materials

Material	Components					Firing Conditions	
-Eucryptite	Alumina	Clay	Talc	Lithium carbonate		Ramps <5 C/min	Hold 1300 C, 3h
Cordierite	Alumina	Silica	Kaolin	Talc	Clay	Ramps <1 C/min	Hold >1400 C, 3h
Aluminum Titanate	Alumina	Titania	Calcium carbonate	Silica	Strontium carbonate	Ramps <1 C/min	Hold >1400 C, 3h

MICROCRACKING UPON THERMAL CYCLING

The microstructure of microcracked materials is shown in Figure 3, for β-eucryptite, cordierite and aluminum titanate, together with that of silicon carbide. The last material does not show any sign of microcracking after the extrusion and firing process.

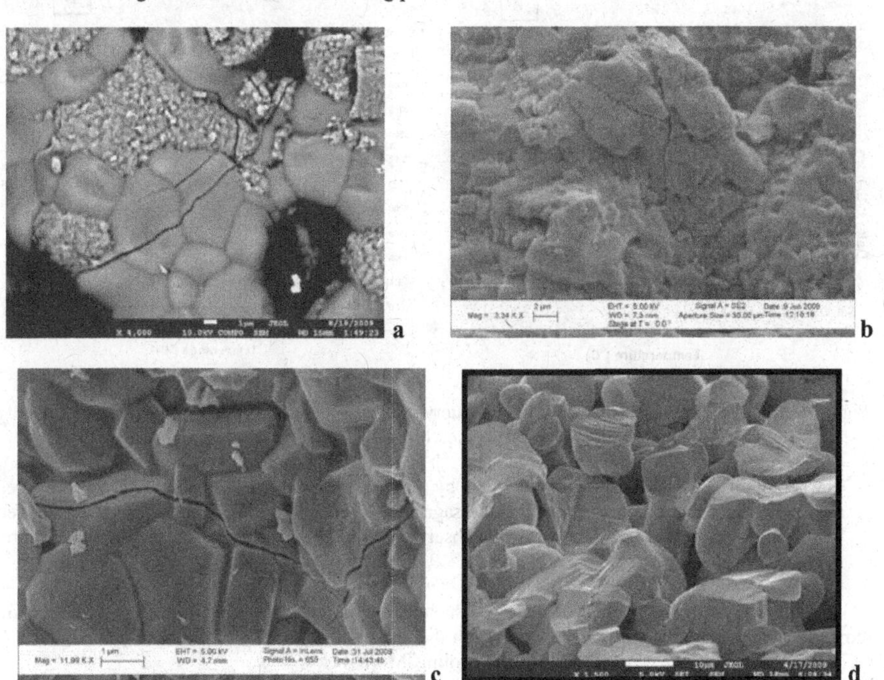

Figure 3. SEM images of β-eucryptite (a), cordierite (b), aluminum titanate (c) and silicon carbide (d). Note the different scales, underlining the different grain and crystallite size.

The consequences of microcracking are multiple:

1- Microcracked materials display a very particular U-shaped expansion curve, as evidenced in Fig. 4.
   a. The coefficient of thermal expansion, CTE, (i.e., the derivative of the expansion with respect to temperature) is negative in a large temperature region. This implies that thermal microcracks close upon heating and open upon cooling, so that the slope of the thermal expansion curve goes through zero. This curve is completely reproducible upon successive thermal cycling.
   b. The expansion follows a hysteresis loop. This implies a retarded microcrack opening upon cooling[10].

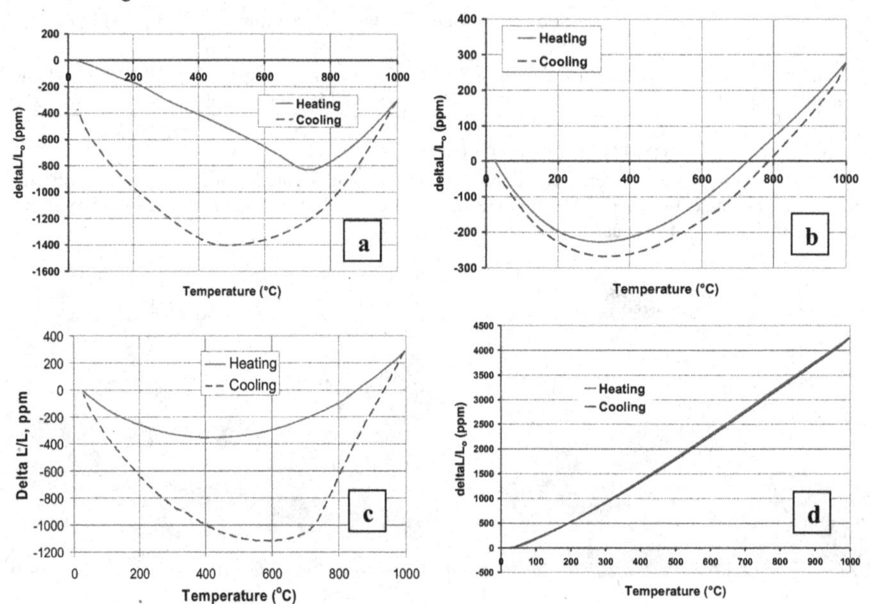

Figure 4. Macroscopic thermal expansion of β-eucryptite (a), cordierite (b), aluminum titanate (c) and silicon carbide (d).

2- The Young's modulus variation, measured by sonic resonance[11] as a function of temperature, shown in Figure 5, conveys the same message and corroborates the explanation given above. Interestingly enough, and in line with what observed for the thermal expansion, microcrack closure renders the materials stiffer.

Relevantly, non-microcracked materials such as SiC do not exhibit any of the features shown above. Indeed, Figures 4d and 5d show that both the thermal expansion and the Young's modulus of SiC cycle on the same curve upon heating and cooling. Moreover:

1- The thermal expansion coefficient is always positive
2- The Young's modulus decreases slowly but steadily as a function of temperature.

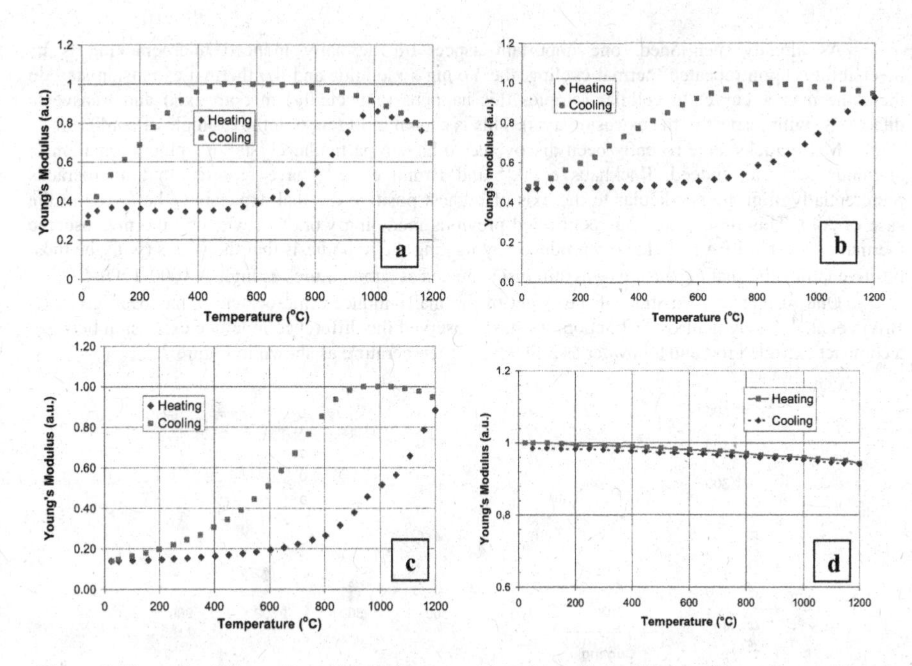

Figure 5. Normalized Young's modulus of β-eucryptite (a), cordierite (b), aluminum titanate (c) and silicon carbide (d), as measured by sonic resonance.

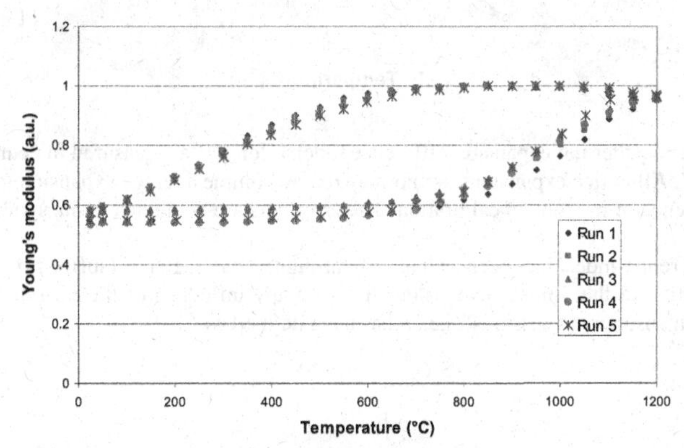

Figure 6. Normalized Young's modulus of cordierite, as measured by sonic resonance, upon successive thermal cycling.

As already mentioned, one important aspect of thermally induced microcracking is its reversibility: upon repeated thermal cycling, the Young's modulus and the thermal expansion stay on the same master curve. In cellular samples this happens (expectedly) in both axial and transverse directions (with respect to the extrusion axis). This is shown in Figure 6 in the example of cordierite.

Microcracks have recently been discovered to have a particular crystallographic orientation in aluminum titanate. Indeed, Backhaus et al.[12] and Bruno et al.[13] have reported that microcracks preferentially align perpendicular to the axis of highest positive thermal expansion (the *b*-axis in the case of AT). This observation has confirmed previous modeling work[13,14], whereby the macroscopic thermal expansion of AT could be reproduced by making the hypothesis that the *b*-axis (with the most positive lattice thermal expansion) was still disconnected at temperatures as high as 1000-1200 C.

One further confirmation of the effect of thermally-induced microcracking has been given by Bruno et al.[14]. Using neutron diffraction, we have observed the difference in lattice expansion between a compact extruded rod and a powder as a function of temperature as shown in Figure 7.

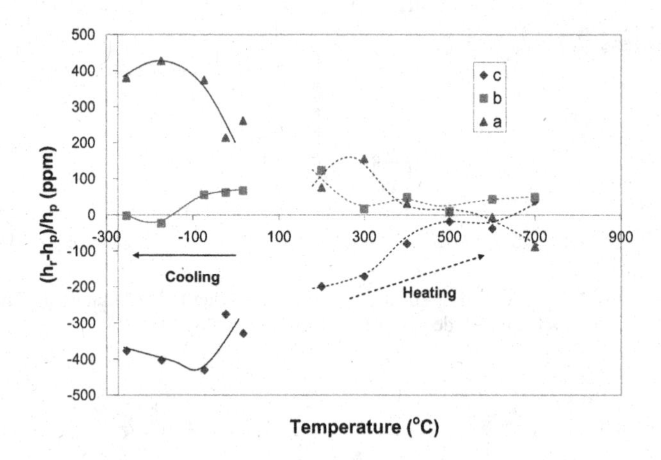

Figure 7. The lattice thermal expansion difference index *h* for AT, as measured in neutron diffraction experiments[14]. All lattice expansions are normalized by volume average expansion to rule out the influence of instrument calibration. The error bars are contained in the symbols.

The difference index has been calculated through a parameter *h* (suffix *r* for rod and *p* for powder), as relative to the volume expansion (to avoid any influence of the instrumental calibration during different measurement sessions), i.e., *h* has been defined as

$$h = \frac{a}{\sqrt[3]{V}} \tag{3}$$

where $V = a \cdot b \cdot c$ is the (orthorhombic lattice) cell volume, and Equation (3) is valid for all crystal axes *a*, *b*, and *c*. Then the difference index has been computed by $(h_r - h_p)/h_p$.

This difference has been taken on cooling from room temperature (RT) to -253 C in the sub-zero range and on heating from RT to 700 C. Assuming the powder particles were single crystals, we considered the index difference as a result of local stresses in the rod where crystallographic expansions are constrained by neighboring grains. The constraint produces the unit cell shape change due to expansion anisotropy (compression on negative CTE axis *c* and tension on positive CTE axes *a* and *b*).

Since the *b*-axis has the highest CTE, it is disconnected from the neighborhood (microcracking) its difference index is insensitive to temperature i.e. the *b* lattice parameter in the rod behaves like its analogous in powder.

Interestingly enough, data in Figure 7 do show that *b* is disconnected, as well as that *c* is under compression and *a* under tension, in line with the model predictions. Figure 7 also shows that upon heating in the high temperature range, we can consider stresses to vanish somewhere above 500 C.

We collected analogous data for cordierite, see[15]: we observed that the measured strain difference between a cordierite compact porous rod and its powder as a function of temperature correspond to the values calculated by the integrity factor model. In the case of cordierite, microcracking is significantly lower than in AT, so that the stress relaxation at temperatures between 500 C and room temperature (RT) is less spectacular. However, we could very well see that the stress-free temperature for cordierite, whereby the lattice expansion is not influenced by internal microstrains, lies above 800 C.

MICROCRACKING UPON MECHANICAL CYCLING

The behavior of microcracked ceramics under applied load presents several very interesting aspects. We have mentioned above that from EBSD (electron back-scattering diffraction) we have observed microcracks in AT being preferentially oriented perpendicular to the *b*-axis (having the most positive expansion). This interesting feature has also been indirectly observed during in-situ neutron diffraction uniaxial compressive testing[13]: the *b*-axis compressive strain is very low even at high applied loads, as shown in Figure 8.

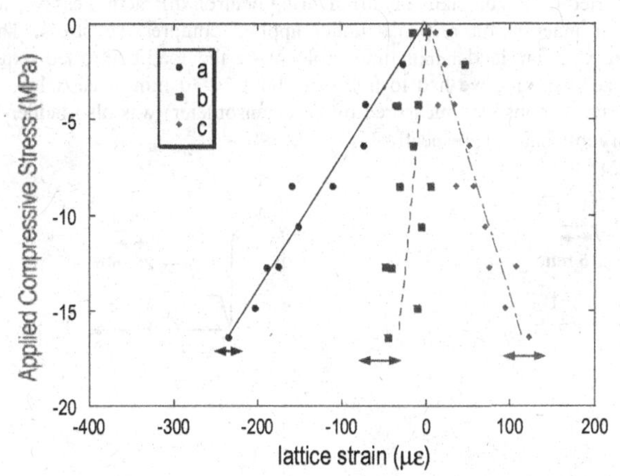

Figure 8. Macroscopic Stress-microscopic (lattice) strain curve for AT. The typical error bars are indicated by arrows. The strains are set to zero for zero applied stress.

The behavior of the *c*-axis (expanding under compressive load) could be explained in the following way: since the *c*-axis has negative lattice expansion[11], the integrity factor model of Efremov[16] would predict that it remains under compression after cooling from the firing temperature[13,14]. Upon the application of a compressive load, the grains oriented with their *c*-axis along the load direction would experience much greater compression and possibly break (see sketch in Fig.9a). This microcracking mechanism would release the compressive residual microstress and the lattice strain would look like in Figure 9b rather than like in Figure 8. In this way the *c*-axis would still be under compression, but part of it would be released by microcracking.

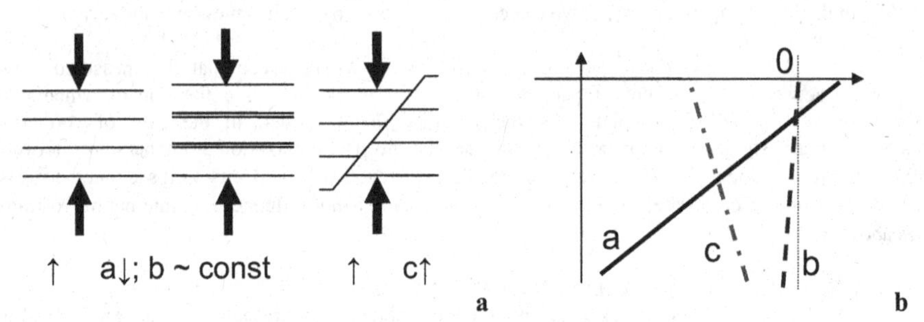

Figure 9. (a) A pictorial view of the mechanism by which the a-axis follows the applied compressive load and the c-axis expands during uniaxial testing. (b) A schematic representation of the data in Figure 8, taking into account the residual microstress in the different crystal directions.

Another important aspect related to microcracking evolution during uniaxial compressive testing has been reported by Pozdnyakova et al.[4]. During neutron diffraction essays, we have observed a strongly non-linear macrostrain response under applied compressive stress. This was possible because of the very peculiar ladder-like time protocol of the load-controlled experiments: during neutron diffraction acquisitions, we had to hold for about 10-30 min at each load (see Figure 10). Surprisingly, the strain response (as measured by an extensometer) was also ladder-like (see Figures 10a and b for β-eucryptite and AT, respectively.

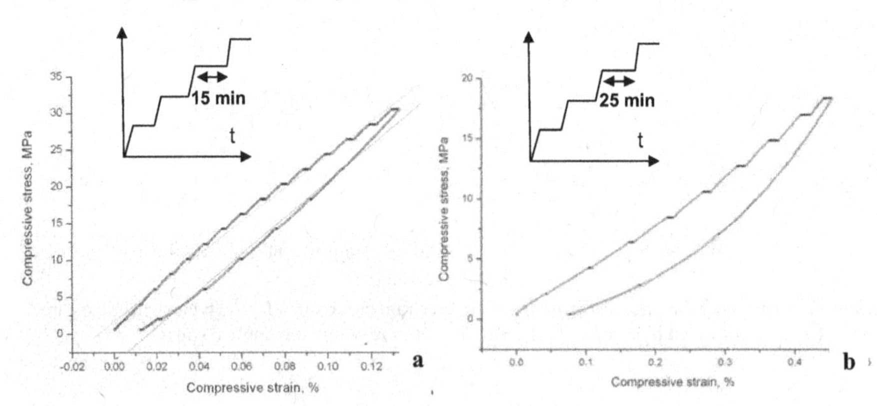

Figure 10. Stress-strain curves for β-eucryptite (a) and aluminum titanate (b) under the stress-time protocol indicated in the insets.

This phenomenon has been associated with a double exponential strain relaxation. The amplitudes A_i and the decay times T_i ($i = 1,2$) of the two exponential functions have shown interesting features (see Figure 11) for both AT and β-eucryptite:

- The first phenomenon has a larger magnitude, increasing with applied load, while the second has lower magnitude and basically constant (for AT it slightly grows); compare Figures 11a and b
- The first phenomenon is much slower than the second for both materials, although they both become slower (increase of T) as a function of applied load, as shown in Figures 11c and d

We will see later that this framework is coherent with thermally induced microcracks being the second type and mechanical microcracks being the first[4].

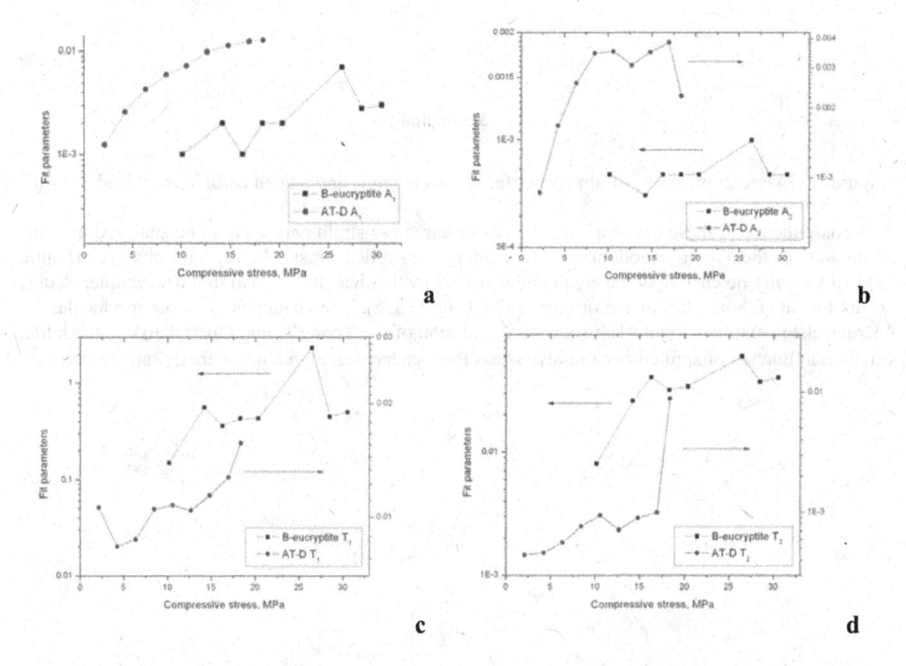

Figure 11. The dependence of the relaxation fit parameters on the applied stress for AT and β-eucryptite during load. (a,b) amplitude, (c,d) decay time.

It is to be noted that for non-microcracked porous materials, such as SiC and alumina, we did not observe any strain relaxation while holding the load during neutron diffraction experiments, although the holding times were shorter (in the range of 5-15 min). However, we could observe some slight non-linearity of the macroscopic stress-strain curves, especially in SiC[17]. This is shown in Figure 12. This non-linearity proved to be associated with a permanent strain at the end of the load/unload cycle, and was therefore explained as due to microstructural damage.

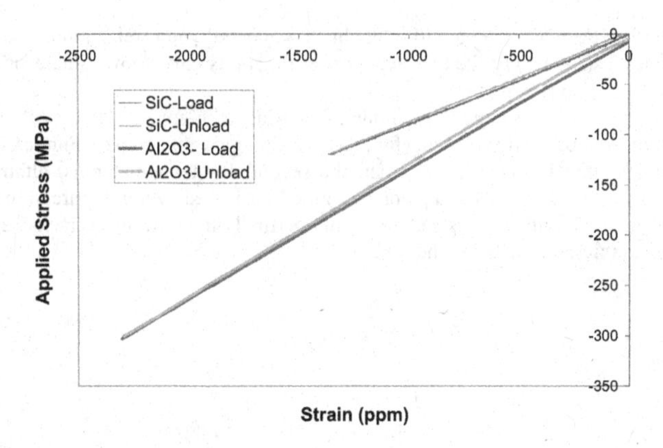

Figure 12. Macroscopic stress-strain curves for porous alumina and silicon carbide upon load/unload.

As a consequence of these observations, the non-linear stress-strain curves could be analyzed in terms of variation of the Young's modulus as a function of the applied stress[4,5,17]. We have observed (Figure 13) that virtually no change of elastic modulus occurs in alumina, and a small, but irreversible, change occurs for SiC (about 20% of the original value (Figure 13a). The change is however spectacular for microcracked ceramics. Figure 13b shows the evolution of the Young's modulus (relative to the initial value) as a function of applied compressive stress for -eucryptite, aluminum titanate, and cordierite.

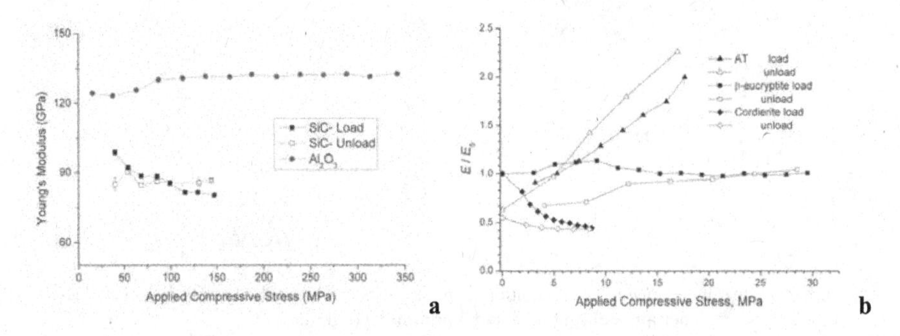

Figure 13. The dependence of the Young's modulus on the applied stress for porous (non-microcracked) SiC and Al_2O_3 (a), and for microcracked AT, cordierite and β-eucryptite (b).

We notice that upon loading β-eucryptite runs through a maximum, aluminum titanate steadily increases and cordierite steadily decreases. Upon unloading, AT follows the same path, cordierite recovers a small part of the initial value, while β-eucryptite runs through a minimum.

Laser Ultrasonic technique allowed us to measure the transverse Young's modulus of the honeycomb specimens under applied axial compression. Results are shown in Figure 14 for AT and cordierite, loaded to about 70% of their strength.

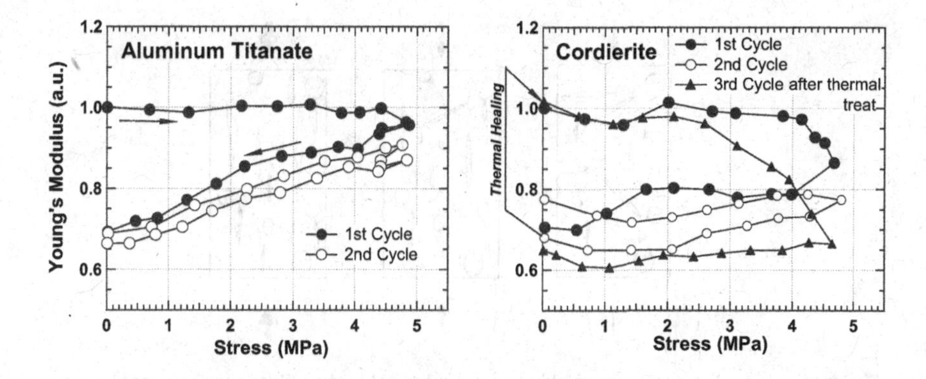

Figure 14. The dependence of the Young's modulus on the applied stress for porous AT and cordierite. Note that the stress values are not corrected for the open frontal area of the honeycomb samples. Subsequent cycles are reported.

Several interesting features can be observed:

1- For AT, the transverse Young's modulus is nearly constant upon unloading, but steadily decreases upon unloading.
2- Upon successive load/unload cycling, it seems to follow the unload path of the first cycle, yet ending slightly lower. The stiffening with applied load is indicative of thermal microcrack closure (contact stiffening).
3- For Cordierite, the transverse Young's modulus is nearly constant up to a certain load and then sharply decreases. Upon unloading, the modulus decreases (or in other words remains at a reduced, damaged, level).
4- Upon successive cycling, the modulus remains almost constant at the reduced, damaged value and may show some additional minor degradation..
5- If, after the first two mechanical cycles, we anneal the samples at 1200 C for 0.5-1h the original stiffness is almost recovered, but the Young's modulus decreases at a lower effective damage threshold than the first cycle and to a slightly greater extent.

DISCUSSION

From the data shown in the preceding sections, we can state that both the macroscopic transverse and axial Young's modulus measurements yield the same scenario: upon mechanical loading (uniaxial compression), thermal microcracks (i.e. preexisting from thermal mismatch effects) close if correctly oriented (perpendicular to the compressive stress or inclined at an angle). However, axial microcracks are generated, at both pore cusps and from existing microcracks (see sketch in Figure 15, and the wing crack model[18]).

Unlike the thermally induced microcracks, the mechanically induced microcracks are not easily healed and a certain amount of them persists at the end of each load/unload cycle. Upon heat treatment, mechanical microcracks can possibly close, but they are more easily re-opened, as indicated by laser ultrasonic experiments. This fact possibly indicates that healing is never complete.

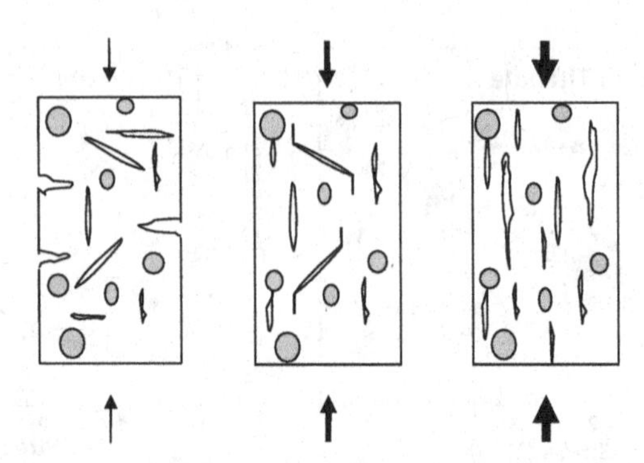

Figure 15. The evolution of microcracking upon mechanical loading. Porosity is sketched by grey circles.

We can therefore rationalize Young's modulus behavior of porous microcracked ceramics as a function of applied load in the following way:

- For β-eucryptite (Figure 13b): at the beginning of loading we start closing existing thermal microcracks, but with increasing applied load the number of mechanical microcracks exceeds the healing phenomenon caused by external compression, and the Young's modulus drops. Upon unloading, thermal microcracks also re-open, and the stiffness ends up being lower than the initial value.
- For AT (Figure 13b), the axial Young's modulus increases steadily because the amount of closing thermal microcracks always exceeds the amount of freshly generated mechanical microcracks. Conversely, the transverse stiffness does not increase, possibly because the amount (or the effect) of closed thermal microcracks (the ones oriented at an angle to the load axis) equals that of newly generated microcracks. Since the latter are mainly axially oriented, upon unloading both the transverse and the axial moduli would decrease, as the thermal microcracks re-open.
- For cordierite (Figure 13b), the amount of thermal microcracks closing with increasing applied load is smaller than the axial mechanical microcracks. Therefore both its axial and transverse stiffness would continuously decrease with load, and remain small after unloading.
- For non-microcracked materials (Figure 13a) the situation is different: SiC and Al_2O_3 can behave similarly to microcracked ceramics, since a certain amount of hysteresis in the macroscopic stress-strain curves is observed. However, the stiffness reduction is absent for alumina and small for SiC, of the order of 20%.

The fact that successive mechanical cycling affects only marginally the stiffness cycle has already been observed by Heap et al.[3] on porous rocks. These authors have seen that acoustic emission (indicating incipient microcracking) would only be triggered if the load in a cycle would exceed the load of any preceding cycle. In this sense, mechanical microcracks are irreversible and do carry the memory of the materials history.

Finally, finite element modeling (FEM) calculations[5,19] have confirmed that convex pore shapes, which are stress concentrators, can be sites for nucleation of new microcracks (see sketch in Figure 16). Expectedly but interestingly, calculations confirm that those microcracks are larger and axially oriented.

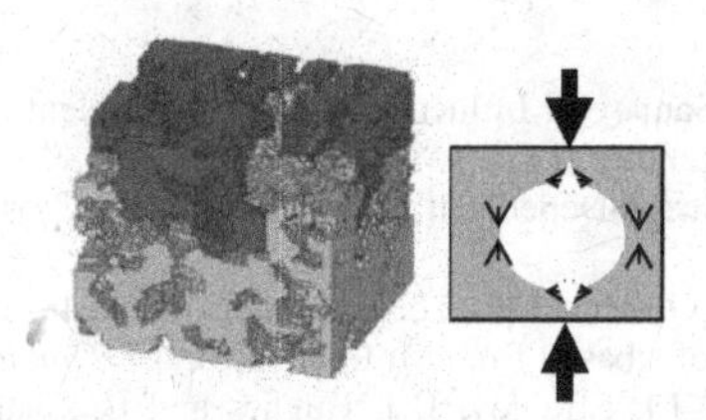

Figure 16. Sketch of the generation of axial microcracks under applied compressive stress, as calculated by FEM. The thick boundary between dark light regions is a long winding microcrack

CONCLUSION

Experimental evidence has been shown that the behavior of microcracked materials under thermal cycling is different from that under mechanical cycling. This has brought us to categorize thermal and mechanical microcracks under two different kinds. In short, we have seen that:

Thermal Microcracks

- Occur because of CTE anisotropy and intergranular mismatch: they nucleate mostly at grain/crystallite boundary
- Build up on cooling from the firing temperature
- Are small and random oriented
- Close and re-open upon cyclic heating/ cooling (reversibility)
- Under applied load they behave as having smaller constant amplitude, but reacting faster to the external field.

Mechanical Microcracks

- Occur because of damage (generate at pore cusps and/or propagate on existing microcracks)
- Build up under applied stress
- Are larger and mainly axial (under compressive load)
- Do not cycle upon load/ unload (irreversibility)
- Under applied compressive load they have larger, growing åmplitude, but they evolve slower than thermal microcracks.

Obviously, the nucleation sites being sometimes common to the two kinds of microcracks, there is no borderline between the two. However, distinct feature have been shown that characterize the two kinds and allow us to better model the materials behavior. Indeed, FEM calculations have well reproduced some of the features we have observed, for both microcracked and non-microcracked materials.

ACKNOWLEDGEMENTS

Andrey Levandovskiy (Corning SNG, St.Petersburg, Russia), Irina Pozdnyakova (CNRS, CEHMTI, Orleans, France), Bryan Wheaton (Corning Inc., Corning, NY, USA), Bjørn Clausen, Don Brown (LANSCE, LANL, Los Alamos, NM, USA), Darren Hughes (ILL, Grenoble, France), Anatoly Balagurov, Valery Simkin, Ivan Bobrikov (FNLP, JINR, Dubna, Russia)

REFERENCES

[1] E.T. Brown, J.W. Bray, F.J. Santarelli, Influence of stress-dependent elastic moduli on stresses and strains around axisymmetric boreholes, *J.Rock Mech.Rock Eng.* **22**,189-196 (1989),

[2] S.Yu, P. Dakoulas, General Stress-Dependent Elastic Moduli for Cross-Anisotropic Soils, *Geotech. Eng.* **119**, 1568-1576 (1993),

[3] M.J. Heap, S.Vinciguerra, P.G. Meredith, The evolution of elastic moduli with increasing crack damage during cyclic stressing of a basalt from Mt.Etna volcano *Tectonophys.* **471**, 153-160 (2009)

[4] I. Pozdnyakova, G. Bruno, A.M. Efremov, D.J. Hughes and B. Clausen, Stress-dependent elastic properties of porous microcracked ceramics, *Adv. Eng. Mater.* **11**, 1023-1029 (2009)

[5] G. Bruno, A.M. Efremov, A.N. Levandovskiy, I. Pozdnyakova, D.J. Hughes, B. Clausen, Thermal and mechanical response of industrial porous ceramics- *Mater.Sci.Forum* **652,** 191-196 (2010)

[6] H.M. Rietveld, A profile refinement method for nuclear and magnetic structures, *J.Appl. Cryst.* **2**, 65-71(1968)

[7] G.L. Squires, Introduction to the Theory of Thermal Neutron Scattering (2nd ed.), Dover Publications, New York (1996).

[8] V. Hauk, Structural and Residual Stress Analysis by Nondestructive Methods: Evaluation - Application – Assessment, Elsevier Science, Amsterdam (1997)

[9] C.B. Scruby and L.E. Drain, Laser Ultrasonics: Techniques and Applications, Adam Hilger, Bristol, Philadelphia, New York (1990)

[10] H.A.J. Thomas, R. Stevens, Aluminium Titanate- A literature review, part 1- Microcracking phenomena, *Br.Ceram.Trans.J.* **88**, 144-151 (1989)

[11] ASTM Standard C623-92, Standard test method for Young's modulus, Shear Modulus, and Poisson's ratio for Glass and Glass-Ceramics by Resonance (2000)

[12] M.Backhaus-Ricoult, C.Glose, P.Tepesch, B.R.Wheaton, J.Zimmermann, Aluminum Titanate Composites for Diesel Particulate Filter Applications, In R.Narayan, P.Colombo. Advances in Bioceramics and Porous Ceramics III, *Ceramic Engineering and Science Proceedings* **31**, Issue 6, 145-156 (2010)

[13] G. Bruno, A.M. Efremov; B.R. Wheaton; J.E. Webb, Microcrack orientation in porous aluminium titanate, *Acta Mater.* **58**, 6649-6655 (2010)

[14] G.Bruno, A.M.Efremov, B.R.Wheaton, I.Bobrikov, V.G.Simkin, S Misture, Micro- and macroscopic thermal expansion of stabilized aluminum titanate, *J.Eur.Ceram.Soc.* **30**, 2555-2562 (2010)

[15] G. Bruno, A.M. Efremov, D.W. Brown, Evidence for and calculation of micro-strain in porous synthetic Cordierite, *Scripta Mater.* **63**, 285–288 (2010)

[16] A.M. Efremov, Impact of domain anisotropy on CTE of isotropic microcrystalline material, *Phil.Mag.* **86**, 5431-5440 (2006)

[17] G. Bruno, A.M. Efremov, A.N. Levandovskiy, B. Clausen, Connecting the Macro and Micro Strain Responses in Technical Porous Ceramics: Modeling and Experimental Validations, *J.Mater.Sci.* **146,** 161-173 (2011)

[18] S. Nemat-Nasser, H.Horii, Rock Failure in Compression, *Int.J.Eng.Sci.* **22**, No. 8-10, 999-1011 (1984)

[19] A.N. Levandovskiy- Private communication (2010)

SIC FOAMS FOR HIGH TEMPERATURE APPLICATIONS

Alberto Ortona*
ICIMSI, SUPSI, Galleria 2, 6928, Manno, Switzerland

Sandro Gianella, Daniele Gaia
Erbicol SA, Viale Pereda 22 P.O. Box 321, 6828, Balerna, Switzerland

ABSTRACT
Silicon carbide open cell ceramic foams with porosity >80% and pore size from 40 to 10 PPI are industrially employed as active zone in porous burners for heat radiation applications.
Si-SiC open cell foams product range is increasing in terms of geometry, foam architecture, and base materials, continuously broadening their fields of application. From the first burners, Si-SiC open cell foams are nowadays employed in catalysis, heat transfer, mechanical and optical applications. This work presents Si-SiC foams main characteristics as well as an overview of their applications in high temperature hostile environments.

SI-SIC CERAMIC FOAMS
Reticulated foams have an open pore structure with an interconnected network of cells, whose edges are made of solid struts (Figure 1). Advanced ceramics are extensively applied as bulk materials for ceramic foams [1]. For high temperature applications, some carbides (e.g. SiC) are, for their outstanding thermo-mechanical properties, the most appropriate. These foams can withstand long oxidative exposing conditions with low material degradation [2],[3].

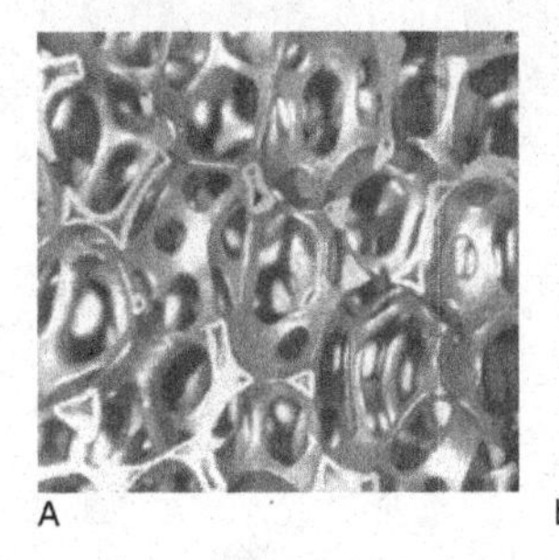
A

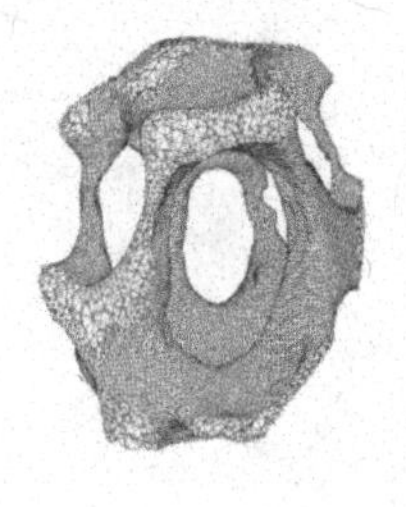
b

Figure 1 a) Si-SiC typical Reticulated Si-SiC foam , b) CT reconstruction of a foam cell.

Ceramic foams are produced mainly through the following methods [4]:

Direct foaming of a liquid slurry
Burn out of fugitive pore formers
Replication of a sacrificial foam template

* Corresponding author : SUPSI-ICIMSI Galleria 2 CH-6928 Manno, tel +41586666611, fax +41586666620, e-mail alberto.ortona@supsi.ch

In the direct foaming of a liquid slurry, a gas is inflated into a suspension or liquid media and subsequently set in order to block bubbles. In most cases, the consolidated foam is sintered at high temperatures to obtain a strengthened porous ceramic.
Burnout of fugitive pore formers, also known as sacrificial template method, consists in the preparation of a ceramic green body with a homogeneously dispersed sacrificial phase which is extracted, generally during heat treatment, to generate pores in the bulk.
Replication technique consists in the impregnation of a cellular structure with a ceramic suspension or liquid polymer precursor in order to produce a macro porous ceramic exhibiting the same morphology of the original porous material [7][5]. These techniques allow the manufacturing of foams with the higher values of porosity and pore size [5].
The most used replication technique is the polymer replica, in which a highly porous polymeric foam is initially soaked into a ceramic suspension, shaken in order to eliminate the excess slip, and pyrolysed by heating between 300 and 800°C. The ceramic coating can be sintered or, in the case of Si-SiC, infiltrated with molten Si. A modification of the polymer replica method consists in pyrolysing a high carbon-yield polymeric foam (e.g phenolic) into vitreous carbon and coating it using Chemical Vapour Deposition (CVD) [6].
Reticulated foams are usually described in terms of a volumetric porosity or by linear pore density (number of Pores Per Inch (PPI)). Foams used in several high temperature applications have porosities in the range 80–90%, and pore densities in the range 5-50 PPI.
ERBISIC foams are produced through the Schwarzwalder method [7]: impregnation of a template, pyrolysis of the polymeric preform followed by liquid silicon infiltration (LSI). The production method allows to obtain a microstructure composed of crystalline SiC and residual silicon with a open porosity smaller than 1%. The resulting material demonstrates high temperature stability up to 1400°C (air), resistance to acid and basic conditions, high thermal conductivity and thermal shock resistance. Its bulk material properties are similar to monolithic Si-SiC composites first produced with the so called "melt infiltration" technique [8].
ERBISIC foams are produced in two varieties, ERBISIC-R, reticulated open cell foams produced starting from a polymer foam, and ERBISIC-F, filamentous foams produced starting with a polymeric wire.

Figure 2 ERBISIC foams product range

Product range is widening in terms of parameters such as macro-porosity, pore size, amount of free silicon; component shape and size can be accurately controlled, as a result of specific customers' needs. Some examples of final products are shown in ***Figure 2***.
A standard 10 PPI foam presents an open macro-porosity of about 87%, a bending strength around 2.5 MPa and high temperature stability up to 1400°C in air. In ***Table 1*** some of their properties are listed.

ERBISIC Foams Properties	
Foam Density [g/cm^3]	0.323
Normalized Density	0.114
Macroporosity [%]	87
Surface Area [m^2/m^3]	550
Av. Strut Thickeness [mm]	0.9
Flexural Strength [MPa]	4
Compression Strength [MPa]	2.5
Thermal Conductivity [W/mK]	8 - 10

Table 1 Some ERBISIC-R (10 PPI) property.

EXAMPLE OF HIGH TEMPERATURE APPLICATIONS

Erbicol developed SiC foams within the European research project BIOFLAM as a component of a liquid biofuels heating system for domestic appliances. SiC foams were then adapted in high energy radiative heating, catalysis, heat transfer, mechanical and optical applications. Beside their well known appliances (e.g. metal filtration), some examples, where a high performance material is needed, are presented below.

Porous burners

Combustion processes are fundamental for the production of energy. To allow combustion to occur, different burner types are available depending on the type of fuel and the operating conditions. Very important is the efficiency of the burner and the reduction of its emissions. Volumetric porous burners can significantly improve these features because this technology allows combustion at comparably low emissions and high surface loads. The combustion is stabilized in an inert open-porous ceramic structure that permits an internal recuperation of heat, a homogeneous temperature field and a compact design. The concept of such a burner is shown in ***Figure 3***. Additional information on the porous burner technology can be found in the literature [9][10][11]; this technology, applied in many different fields of energy conversion [12][13][14], is strongly dependent on high-temperature-resistant porous components.

The porous combustion offers exceptional advantages compared to techniques of free flame burners. This relatively new technology is characterized by higher burning rates, increased flame stability with low noise emissions and lower combustion zone temperatures which lead to a reduction in NO_x formation. Porous burners also allow low CO emissions. Additionally, complex combustion chamber geometries, which are not feasible with conventional state of the art combustion techniques, are possible.

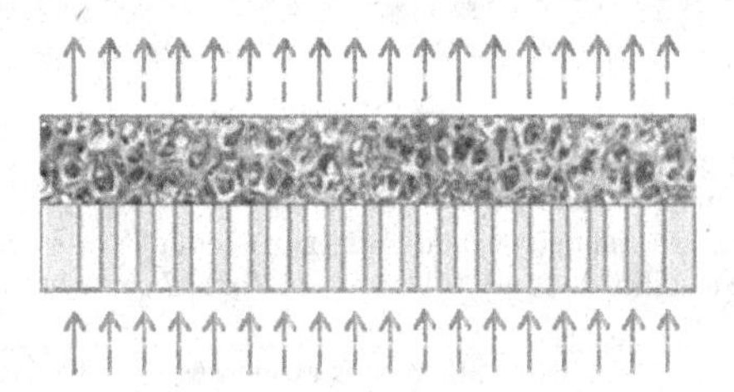

Figure 3 Porous burner for infrared heating. A) working scheme, B) compact system (source LSTM).

Reformers

In Solid Oxide Fuel Cell (SOFC) systems, synthesis gas (syngas) can be produced from hydrocarbon fuels to be directly utilized. Syngas is obtained from liquid fuels (e.g. heating oil, gasoline, diesel) and gaseous fuels (e.g. natural gas, LPG) by steam reforming, catalytic or thermal partial oxidation (POX) and autothermal oxidation. Compared with the other reforming technologies the exothermic thermal POX process [16] has no need for external heat sources and additional feeds like water as in steam reforming. Moreover, the process is catalyst free avoiding catalyst deactivation.

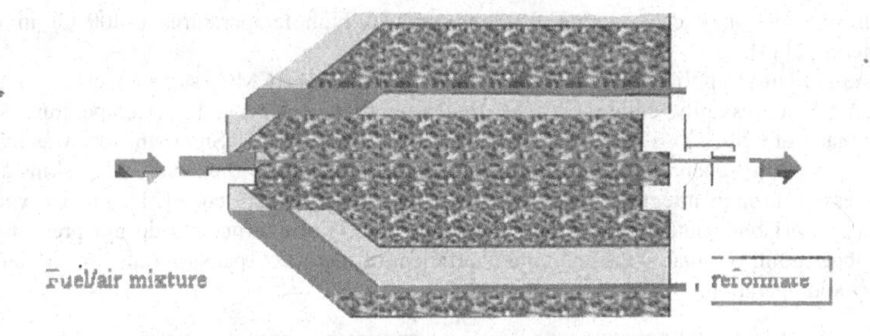

Figure 4 A schematic representation of the TPOX reformer [16]

In this application it was developed a system for the production of syngas under the European project FlameSOFC. A multi-fuel, modular SOFC micro Combined Heat and Power (CHP) was developed, which is capable to operate with gaseous and liquid hydrocarbon fuels [17]. A special reactor [16] was built, based on porous media, with a conical and a cylindrical section (***Figure 4***). In the diverging conical section the oxidation zone is stabilized. Downstream in the cylindrical section reforming reactions take place. The process is performed with preheated air (up to 700° C) premixed with methane just before entering the porous reaction zone. The drawback of this solution is soot formation. Different ceramic materials were employed into the porous reaction zone. Higher heat recuperation was detected in the case of SiC foam based reformer. Regarding the soot point, the SiC foam based reactor showed a better performance than the other ceramics.

Catalytic support

Si-SiC open cell foams are successfully applied as catalyst support [18]. Their open-cell ceramic structures can be fabricated in a variety of shapes and allows high porosities with material interconnectivity [15]. These characteristics lead to a very low pressure drop, if compared with packed beds, and a high contribution of convection heat transfer inside the macro pores. Due to their morphology foams can be easily coated with high-surface-area catalytic micro porous layers Many applications involving important catalytic reactions have appeared in the open and patent literature, especially for processes which need low pressure drop with low-contact-time reactions at high space velocities or with narrow reactors in heat-transfer-limited systems and in controlling axial and radial temperature profiles in highly exothermic and endothermic reactions.
Furthermore Si-SiC foams are mostly suitable in strong acid or basic conditions, exothermic reactions and reactions for which metal or oxide ceramics catalyst carriers are not suitable.
An interesting application, which somehow is linked to others described below, is the solar reforming. The concept is to use an endothermic reaction to absorb heat from concentrated solar light that can be stored and later used to generate heat for several chemical reactions aiming to produce Hydrogen [18].

Structural application

Sandwich structures are very effective because of their high specific stiffness and strength; ceramic based sandwiches can enhance performances of hot structures or thermal protection systems. Such structures can be manufactured in two sets of elements: some with mechanical functions (shell, fasteners, and stand-offs), and others with thermal functions (inner insulation layers, seals and insulating washers).
Reticulated SiC foams can bear high thermal loads, high thermal shocks [3] and, due their high porosity (> 80%), they have rather low effective thermal conductivities [19]. If passive oxidation

conditions are met they can operate for long time at high temperatures (1400°C) in oxidative environments [2] [3].

For thermo-structural applications, Ceramic Matrix Composites (CMC) skins together with ceramic foams can be successfully employed. The first attempt to realize a high temperature structured sandwich made of CMC skins and a Chemical Vapor Infiltration (CVI) SiC foam core was reported by Fisher in 1985 [20]. Recently NASA developed a sandwich made with C_f-SiC_f/SiC_m skins and a SiC core, processed through integral densification of the CMC and foam core [21]. In this work SiC is deposited by CVI both onto fibers preforms and the core. These structures do not present a joining layer. Authors point out that skins and core coefficient of thermal expansion must be similar both for processing and operation.

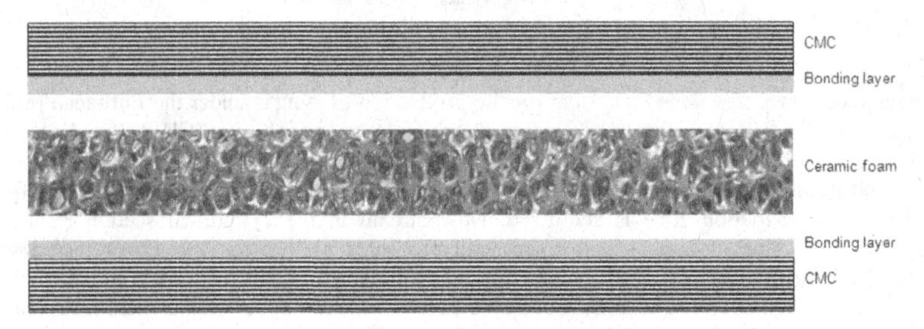

Figure 5 Sandwich assembly

A procedure to assemble ceramic sandwich structures in any shape with low-cost manufacturing techniques usually employed in the polymer matrix composites fabrication was recently presented [22]. Sandwich skins and core are assembled at the very beginning of the sandwich manufacturing employing a polymer derived bonding layer, the assembled sandwich structured composite "preform" can be further densified with the common techniques employed for CMC manufacturing. A first attempt has been made via Polymer Impregnation and Pyrolysis (PIP). In these sandwiches, load is transferred from the faces to the core, not by adhesion but, as foams struts are embedded into the bonding layer, by mechanical joining. Upon bending these sandwiches presented a marked toughening behavior

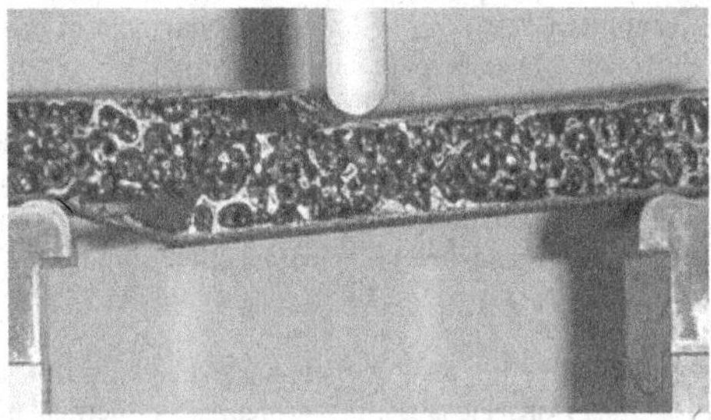

Figure 6 Failure mode of the ceramic core CMC skins sandwich

Bonding is strictly related to skins and core coefficient of thermal expansion matching. In bonded sandwich ultimate strengths is higher than that of the equivalent plain foam. After reaching a maximum stress all samples did not experience a catastrophic failure. In this sandwich configuration failure was observed into the ceramic core mainly because of shear.

Solar radiation absorbers

Solar tower technology is a promising way to generate large amounts of electricity from concentrated solar light The concentrated radiation is generated by many mirrors which reflect the solar radiation onto a focal point known as "solar air receiver" which absorbs the radiation and converts it into high temperature heat [23].

Some concentrated solar systems are made by a receiver, also called volumetric air receiver, in which air passes through a porous material which absorbs the concentrated solar radiation ***Figure*** *7* at the same time.

Ambient air enters the material from the side facing the concentrated solar energy absorber, keeping it relatively cold. In principle, the temperature distribution should be as per ***Figure*** *7*. With still cold air at the front minimizing thus thermal radiation losses.Going through the porous body the air temperature increases till reaching an equilibrium after a couple of cell diameters.

The requirement for the material in the solar absorber applications are temperature resistance up to 1200°C, thermal shock resistance (>200 K/min air outlet temperature), thermal cycle resistance, high thermal conductivity, low pressure drop and obviously a low reflectivity. Further requirements are lower cell size to achieve large surface areas necessary to transfer heat from the material to the gas. Among different solutions adopted for porous bodies are a ceramic or metallic fiber mesh materials, siliconized silicon carbide (SiSiC) catalyst carrier with parallel channels and reticulated SiC foams.

SiC foams are very promising because porosity can be tuned in order to optimize volumetric absorber performances and also because they allow a three dimensional flow, avoiding stagnation points which lead to material overheating and failure.

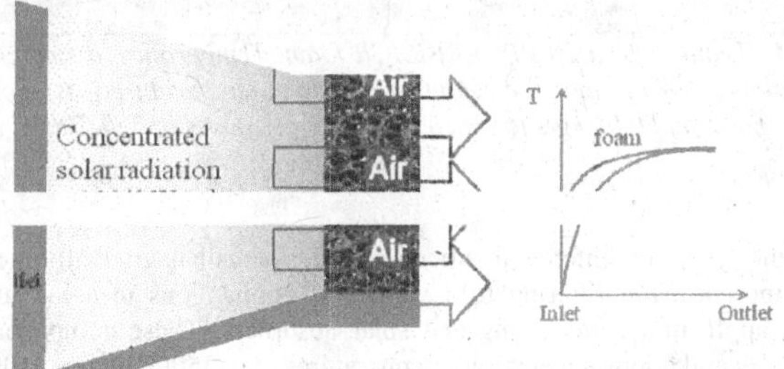

Figure 7 The volumetric solar receiver principle [23]

The inner surface of the foams can be coated with catalytic materials when serving a combined absorber/reactor in a solar chemical process. The Si-SiC foams shows high solar thermal efficiency, high thermo-mechanical strength and chemical stability.

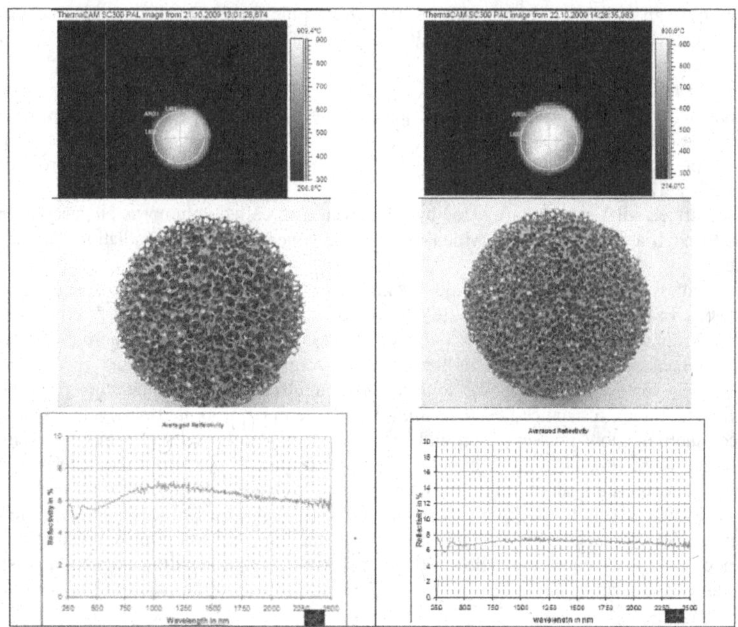

Figure 8 Left a 10 PPI and right a 20 PPI ERBISICR foam. Temperature distribution and reflectivity with irradiation of about 3kW on aperture (~600kW/m²). Inlet temperature was about 45°C. Mass flow was varied between 0.003 and 0.01 kg/s to reach outlet temperatures of 250-700°C (courtesy of DLR)

CONCLUSIONS

During the past twenty years the interest in silicon carbide foams has gradually increased, and moved from the classical molten-metal filtering to a variety of applications in areas such as combustion, catalysis, structural applications, reforming and solar absorbers. These components must withstand repeated thermal cycles and a long time at high temperatures (e.g. 35000 hours, >1000°C).
Optimization of the thermal and mechanical properties was Erbicol's main activity during the last ten years. From the first foams developed for porous combustors, the ability to produce ad-hoc foams has opened the horizons for other applications, some of which are in phase of industrialization, and others requiring further research.

REFERENCES

[1] M. Scheffler and P. Colombo eds. "Cellular Ceramics: Structure, Manufacturing, Properties and Applications,", WILEY-VCH Verlag GmbH, Weinheim, Germany, 2005

[2] *R. A. Mach, F. v. Issendorff, A. Delgado A. Ortona,* Experimental investigation of the oxidation behavior of Si-SiC-foams *F. 32nd* Advances in Bioceramics and Porous Ceramics, 299-311,2009

[3] A. Ortona, S. Pusterla, P. Fino, F. R. A. Mach, A. Delgado, S. Biamino, "Aging of reticulated

Si-SiC foams in porous burners" Advances in Applied Ceramics, 2010,Vol. 109, n°4, pp 246-251

[4] Colombo Conventional and novel processing methods for cellular ceramics Phil. Trans. R. Soc. A January 15, 2006 364:109-124

[5] A.R. Studart, U. Gonzenbach, E. Tervoort, L.J. Gauckler "Processing Routes to Macroporous Ceramics: A Review" J. Am. Ceram. Soc., 89 [6] 1771–1789 (2006)

[6] Sherman, A. J., Tuffias, R. H. & Kaplan, R. B. 1991 Refractory ceramic foams: a novel, new hightemperature structure. Am. Ceram. Soc. Bull. 70, 1025–1029

[7] K. Schwartzwalder, A. V. Somers: Method of making porous ceramic articles, US patent No. 3090094, 1963

[8] Hillig W.B., Mehan R.L. Morelock C.R. DeCarlo V.J. Laskow W., Silicon/Silicon Carbide Composites, Ceramic Bullettin, 1975, vol. 54, No 12, 1054-1056

[9] D. Trims, F. Durst, Combustion in porous medium - advances and applications, Combust. Sci. Technol., 121, 153-168 ,1996.

[10] D. Trims, Porenbrennertechnologie – Ein Überblick, GWF, 2, 92-99 , 2006.

[11] S. Wood, A.T. Harris, Porous burners for lean-burn applications, Progress in Energy and Combustion Science, 34-5, 667-684, 2008

[12] S. Diezinger, P. Talukdar, F. v. Issendorff, D. Trimis, Verbrennung von niederkalorischen Gasen in Porenbrennern, Gaswärme Int., 3, 197-192, 2005.

[13] M.M. Kamal, A.A. Mohamad, Combustion in porous media, Proc. Inst. Mech. Eng. Part A: J. Power Energy, 5, 487-508, 2006.

[14] M. Abdul Mujeebu, M.Z. Abdullah, M.Z. Abu Bakar, A.A. Mohamad, M.K. Abdullah, Applications of porous media combustion technology - A review, Applied Energy, Volume 86, Issue 9, September 2009, Pages 1365-1375

[15] J. Grosse, B. Dietrich, G. Incera Garrido, P. Habisreuther, N. Zarzalis, H. Martin, M. Kind, Be. Kraushaar-Czarnetzki, Morphological Characterization of Ceramic Sponges for Applications in Chemical Engineering, Industrial & Engineering Chemistry Research 2009 48 (23), 10395-10401

[16] Z. Al-Hamamrea, S. Voß, D. Trimis, "Hydrogen production by thermal partial oxidation of hydrocarbon fuels in porous media based reformer", international journal of hydrogen energy 34 (2009) 827–832.

[17] S. Voss, D. Trimis, J. Valldorf, "Development of a Fuel Flexible, Air-regulated, Modular, and Electrically Integrated SOFC-System (FlameSOFC)", 18th World Hydrogen Energy Conference 2010 - WHEC 2010,Parallel Sessions Book 6: Stationary Applications / Transportation Applications, 2010, Essen

[18] Twigg, M.V., Richardson J.T., Fundamentals and applications of structured ceramic foam catalysts Ind. & Eng. Chem. Res., 2007, VL 46, 4166-4177

[19] A. Ortona ,S. Pusterla ,S. Valton , "Reticulated SiC foam X-ray CT, meshing, and simulation", Advances in Bioceramics and Porous Ceramics III, Ceramic Engineering and Science Proceedings, 2010, Volume 31, Issue 6.

[20] Fisher R. Burkland C. Busmante W., Ceramic composite thermal protection systems, Ceram. Eng. Sci. Proc. Vol. 6, no. 7/8, pp. 806, 1985

[21] Hurwitz F. I., Steffier W., Koenig. Kiser J., Improved Fabrication of Ceramic Matrix Composite/Foam Core Integrated Structures, NASA Tech Briefs, 2009, 36.

[22] A. Ortona, S. Pusterla, S. Gianella "Cf/SiCm CMC - SiC foam sandwich preparation and characterization" Conference proceedings HTCMC7, Bayreuth (D), 2010

[23] Thomas Fend, "High porosity materials as volumetric receivers for solar energetic", Optica Applicata 2010(Vol.40), No.2, pp. 271-284.

POROUS SiC CERAMIC FROM WOOD CHARCOAL

S. Manocha*, Hemang Patel, and L.M. Manocha

Department of Materials Science
Sardar Patel University, Vallabh Vidyanagar-388120 Gujarat, INDIA
*Corresponding Author: Email: sm_manocha@rediffmail.com

ABSTRACT

Development of Bio-SiC materials with tailor made microstructure and properties using natural biopolymeric cellulose template are of great current interest. Porous β-SiC ceramic has been fabricated by different methods using wood charcoal infiltrated with silica sol as well as single step method by using pine wood powder. Pine wood charcoal impregnated with resin and subsequent carbothermal reduction process at 1650^{0}C in inert Ar atmosphere produced light weight biomorphic β-SiC ceramic with good mechanical properties and porosities. The highly anisotropic cellulose structure of pine wood generates novel cellulose ceramics with a meso and macro porous structure pseudomorphous to the initial porous tissue skeleton. Microstructural observation and phase identification of resulting wood ceramics have been performed by scanning electron microscopy (SEM) and X-ray diffraction (XRD), respectively. Weight loss during heating of wood and phenolic resin in N_2 atmosphere was investigated by thermogravimetric analysis (TGA). Experimental results showed that porous β-SiC possessed topologically homogenous pore structure. This provides a low- cost and ecofriendly route to advanced ceramic materials, with near-net shape potential.

1. INTRODUCTION

In the recent years the interest in porous materials with cellular structures, such as foams, reticulated and biomorphic materials has increased due to their specific properties such as low density, low thermal conductivity, thermal stability, high-surface area, and high permeability [1-3] These properties make porous ceramics suitable for a wide range of technological applications, including, catalyst supports, filters for molten-metals and hot gases, thermal insulators, refractory linings, and biomaterials [4,5] These biopolymers give rise to three-dimensional structures. Because of these unique structures, woods exhibit a remarkable combination of high strength, stiffness, and toughness. Since these properties are retained after pyrolysis of wood. Therefore, low cost biomorphic composites having excellent mechanical properties are produced. [6]

Various processing routes have been proposed for production of porous ceramics, including polymeric sponge [7-9] and direct foaming with different foaming agents. [10, 11] The processing parameters of fabrication determines the range of porosity, the pore size distribution, and the pore morphology. The polymeric sponge method [7] is a simple, inexpensive and versatile way for producing ceramic foams. The method consists in the impregnation of a polymeric sponge with slurries containing appropriate binders, followed by a heat treatment to burn out the organic template (foam) and to sinter the remaining skeleton. The anatomical features of naturally grown materials are used as template for the design of porous cellular ceramics and ceramic composites, and the morphology of the native tissue is maintained in the final ceramic product. [12-14] The main innovative characteristic of this methodology is the possibility to design macro- and microcellular parts, which could not be produced by other conventional techniques. In the recent years, different biotemplating technologies were developed for the conversion of biological templates such as wood, cellulosic and fibrous materials into biomorphic ceramics. [14]

A number of these fabrication approaches have utilized natural wood or cellulosic fibers to produce carbon preforms. [15-17] Representative fabrication processes of biomorphic composites are liquid silicon infiltration (LSI), silicon vapor infiltration, and chemical vapor infiltration and reaction (CVI-R) processes. Silicon vapor infiltration, and chemical vapor infiltration and reaction processes are used to produce high porous SiC ceramics based on biological structures. Due to the low temperature and silicon content involved in these processes, the growth rates are very slow (microns per hour) and the microstructure of the obtained composites is very porous, thus limiting their use in producing bulk ceramics. However, a significant increase of the synthesis rates can be achieved when cellular biological structures with an open porosity are used. [18-24] Good wetting of the silicon melt to carbon lead to a rapid infiltration of silicon and fast reaction to form SiC. The reaction rate between the carbon and the silicon was faster than the infiltration rate, the higher the initial pore volume fraction of the preform, the greater the amount of free silicon in the reaction product. The infiltration of molten silicon into carbon preforms from wood, which consists of a carbon network of open channels, could produce biomorphic C/Si/SiC composites. Carbon preforms have a large volume fraction of pores. Hence, the molten silicon will react with the carbon preforms to form SiC, and the excess silicon could fill the open channels in the carbon preforms. The use of wood provides a low-cost starting material that has near-net and complex shape capabilities, in contrast with the simple shapes that are normally produced by traditional ceramic processing techniques. An increased shape capability, using readily machinable carbon preforms, could reduce the processing and machining costs significantly. [25, 26] The main processing methods used for the preparation of cellular ceramics such as foams, honeycomb structures and interconnected rods, fibers and hollow spheres are recently reviewed by Colombo. [27] The fabrication and use of cellular ceramics in a vast number of different fields has also been described in a recent book edited by Scheffler and Colombo. [28] While this valuable literature contains extensive information on the production and applications of cellular ceramics, the processing– microstructure–property relations for each of the main processing routes has not been fully explored. [29]

Although understanding of these relations would greatly aid the selection of processing techniques that can provide the final microstructure and properties required for each specific application. In the present work two methods involving single step has been developed for the formation of silicon carbide.

2. EXPERIMENTAL

2.1 Sample Preparation

Bio-SiC ceramics were prepared from two different techniques. In the first technique powder route and in the second solution route was used. In the powder route, the pine wood was powdered in the ball-mill to a mean size 75μm. The wood powder was mixed with phenolic resin and silica powder in different molar ratios by using ball mills. The mixing was done for 12hrs. The powder was hot pressed to make pellets in the hydraulic press. The pressure was maintained up to $150 kg/cm^2$. The pellets were heated up 1650^0C for 4hr in inert Ar atmosphere. In solution route pine wood powder of 75 μm size was mixed with liquid phenolic resin and silica powder having mean size ≤75 μm in different molar ratio. The solution was poured in to moulds at room temperature for 12hr and at 100^0C for 24 hr. After 24hr the dried pellets were heated to 1650^0 C for 4hr in Ar atmosphere. The SiC made from these two techniques were characterized. The flow chart for the preparation process of porous SiC is shown in fig.1 (a) & (b)

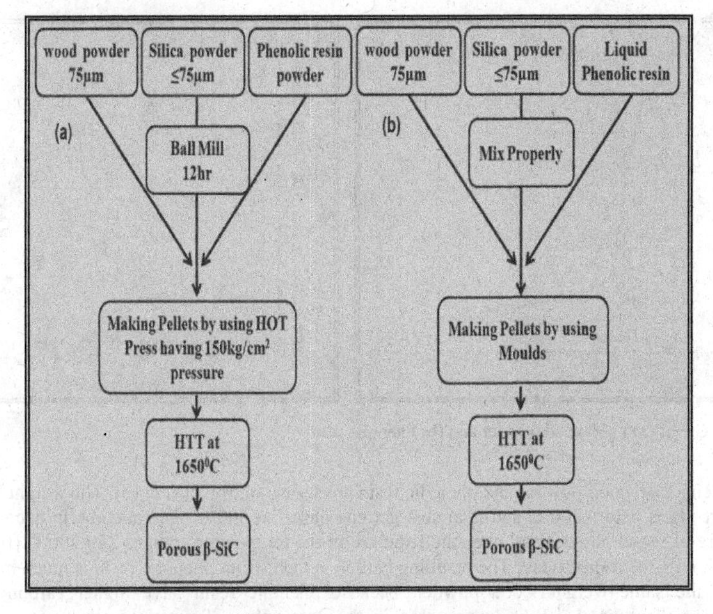

Fig.1 (a) Flow chart of preparation of porous SiC through powder route

(b) Flow chart of preparation of porous SiC through powder route

2.2 Characterization

TGA (Mettler TG-50) analysis was carried out to determine the temperatures of pyrolytic degradation of pine wood and phenolic resin. Scanning electron microscopy (SEM Hitachi S-3000N) was used to characterize and analyze the microstructure of wood ceramic and porous SiC ceramic. X-ray diffraction (Philips, X`pert model) was employed for phase identification and to examine the crystallinity of resulting SiC ceramic. BET (Micromeritics, Gemini-2375) surface area analysis was used to determine surface area, average pore diameter and total pore volume of resulting porous SiC ceramics.

3. RESULTS AND DISCUSSION

3.1 TGA analysis of wood powder and pine wood

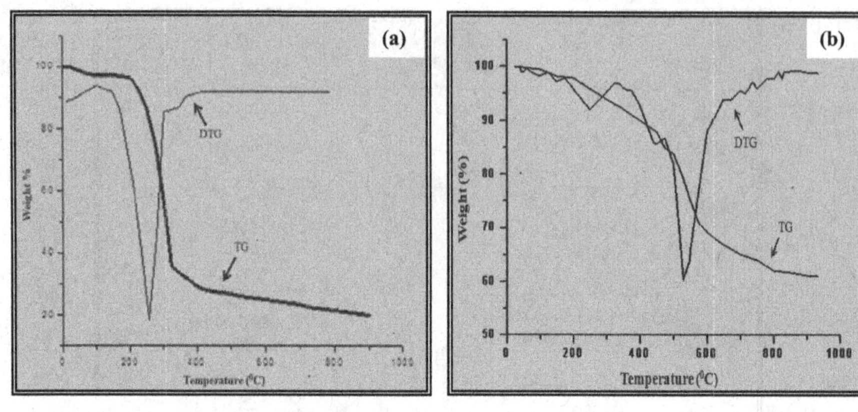

Fig.2 TGA of (a) Pine wood powder and (b) Phenolic resin

The TGA curve of wood powder and phenolic resin are shown in Fig. 2(a) & (b). The weight loss in N_2 atmosphere starts below 200^{0}C and is almost get completed at 600^{0}C. The maxima in decomposition temperature of wood powder and phenolic resin are in the temperature ranges 240-400^{0}C (fig.2a) and 500-600^{0}C, (fig.2b) respectively. The resulting carbon content from phenolic resin is much higher than that from the same weight wood powder, because phenolic resin have higher carbon yield on carbonization than wood.[30] It is known that in the conversion of wood powder into carbon, the following stages are included: (a) desorption of adsorbed water up to 150^{0}C, (b) splitting off of cellulose structure water between 150 and 240^{0}C [31], (c) hydrocarbon structure formed through the chain scissions, or depolymerization, and breakdown of C-O and C-C bonds within ring units evolving water, CO and CO_2 between 240 and 400^{0}C, (d) aromatic polynuclear structures started to form above 400^{0}C and developed gradually above 500^{0}C [32], (e) aromatic reactions occurred and the carbon network shrunk to accommodate the excessive volume left by the evolving gases between 400 and 800^{0}C, (f) thermal induced decomposition and rearrangement reactions were almost terminated leaving a carbon structure above 800^{0}C. As seen in the TG curve of phenolic resin, (Fig.2b) weight loss occurs at about 200^{0}C, and rapidly develops above 450^{0}C, indicating the formation of a great deal of gas. Weight loss rate reaches maxima at approximately 550^{0}C, and evidently decreases above 600^{0}C.

According to Yamashita et al.[33] thermal weight loss below 450^{0}C was caused by the dehydration reaction of phenolic resin, including thermo curing reaction between hydroxyl methyl groups and hydrogen groups within aromatic rings and the condensation reaction between methylene and hydroxyl groups. The condensation aromatic polynuclear structure started to form above 450^{0}C and developed above 500^{0}C [34], releasing small molecular substances such as CH_4, H_2, CO, CO_2, etc. Weight loss at approximately 700^{0}C was attributed to further carbonization and dehydrogenation reactions in phenolic resin. As a result, uniting phenol compound advanced and the dehydration and depolymerization reaction were promoted. Wood and phenolic resin were converted into amorphous carbon and glassy carbon after carbonization, respectively.

3.2 Surface Area Analysis by BET method

Surface area of both biomorphic SiC was determined by BET method. Fig 3 (a) and 3 (b) shows type IV adsorption isotherm porous SiC made from powder and solution route respectively. Type IV isotherm possess a hysteresis loop, the shape of which varies from one adsorption system to another. Hysteresis loops are associated with meso porous solids where capillary condensation occurs. [35] BET surface area of SiC made from powder route is $3.7288 m^2/g$ and the average pore diameter by BJH method is 22nm. While the BET surface area of SiC made from solution route is $1.3367 m^2/g$ and in this case the average pore diameter was not determined by BJH method may be due to the very loose and fluffy structure of SiC.

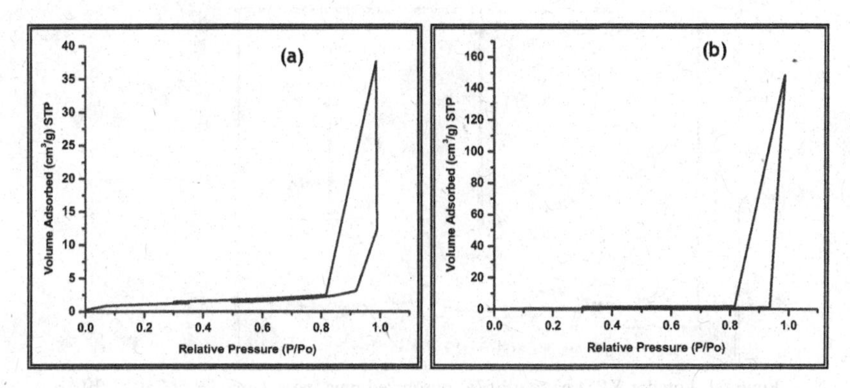

Fig.3. Nitrogen adsorption isotherm of (a) SiC made from powder route (b) SiC made from solution route

The porosity of as such green pellets and final SiC ceramics obtained from powder and solution routs are shown in table I.

Table I. Kerosene porosity of pellets obtained from powder and solution route.

Sr.No	Sample	Kerosene Porosity (Powder Route)	Kerosene Porosity (Solution Route)
1	Green Pellets	1.33%	36.77%
2	SiC ceramic	82.78%	Not detected due to fluffy nature of SiC

3.3 XRD Analysis

The XRD pattern of pyrolysed pine wood powder and phenolic resin shows similar results. Carbon obtained from both pine wood and phenolic resin was amorphous in nature. It can be seen that two broad peaks centered around 23^0 (0002) and 44^0 (0004), suggested the amorphous nature of carbon. XRD of the SiC shows the sharp peaks and the substantial amount of silica phase can be detected at 43^0. One additional line at $2\theta = 33.67°$ (d= 0.260nm) in the XRD pattern of the porous SiC ceramic is also detected near the (111) line of the cubic structure of SiC at $2\theta = 35.60°$ (d111 =0.252nm) which is characteristic of the hexagonal polytypes [α-SiC phase] [36]

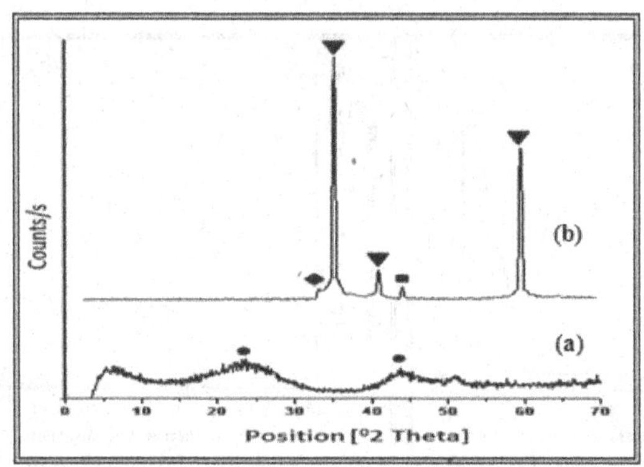

Figure 4. Powder XRD pattern of (a) pyrolyzed pine wood powder at 750°C (b) SiC ceramic

Where: ▼= β-SiC,◆= α-SiC, ■= Crystobalite, •= Carbon

It can be seen that β-SiC ceramic prepared by these two techniques contains small amount of silica (crystobalite phase). Excess amount of free silica can be removed by HF leaching. Table II shows the percentage silica content in final SiC ceramic.

Table II. Percentage silica content in SiC ceramic.

ND= Not detected

Sr.No	Sample (SiC)	%C	%SiO_2	%SiC
1	Solution route	ND	14.28	85%
2	Powder route	ND	12.03	87.7%

3.4 Mechanism for the growth of SiC whisker and SiC particles

It has been shown [37] that under the experimental conditions of the present work, the overall reaction between charcoal and silica for producing silicon carbide are:

$$SiO_{2\,(s,l)} + 3C_{(s)} \longrightarrow SiC_{(s)} + 2\,CO_{(g)} \quad \text{................1}$$

Where s, l and g refer to the solid state, the liquid state and the gas state, respectively. In fact, reaction (1) proceeds through two stages in which a gaseous intermediate, silicon monoxide (SiO) gas is formed. The first step consists of a solid-solid or solid-liquid type of reaction between carbon and silica leading to the formation of gaseous silicon monoxide (SiO) and carbon monoxide (CO) according to reaction (2)

$$SiO_{2\,(s,l)} + C_{(s)} \longrightarrow SiO_{(g)} + CO_{(g)} \quad \text{................2}$$

Eq. (2) is either a purely solid–solid or a liquid–solid reaction (above 1450^0 C, quartz melts).[38] Once carbon monoxide (CO) is formed, SiO can also be produced according to reaction.

$$SiO_{2\,(s)} + CO_{(g)} \longrightarrow SiO\,(g) + CO_{2\,(g)} \quad \text{................3}$$

Since a significant amount of carbon remains in the material, any CO_2 produced will be consumed immediately by the Boudouard reaction (4) to form CO gas

$$CO_{2\,(g)} + C_{(s)} \longrightarrow 2CO_{(g)} \quad \text{................4}$$

In these reactions, carbon is either a CO_2 getter or a CO generator, which keeps the CO_2/CO ratio low enough to make the reduction of SiO_2 possible by gas phase CO. In second step, the gaseous silicon monoxide (SiO) subsequently reacts further with carbon according to the following gas-solid reaction:

$$SiO_{(g)} + 2C_{(s)} \longrightarrow SiC_{(s)} + CO_{(g)} \quad \text{................5}$$

The SiO vapor from Eqs. (2) and (3) reacts with carbon to yield SiC(s) nuclei heterogeneously on the surfaces of carbon through Eq. (5), which is commonly accepted mechanism of bulk SiC formation.[39] The synthesized SiC is strongly dependent of the carbon source,[40] which makes it possible that the resulting SiC ceramic is of a woodlike microstructure. In general, as soon as SiC forms on carbon, the growth process via Eq. (5) can be hindered by either the solid diffusion of carbon or the diffusion of SiO gas molecules through SiC layer.

Under SEM investigations, it is noted that SiC whiskers exists (or fibers) in solution route, as shown in fig. 6. The formation of SiC whisker/fiber cannot be explained by the gas–solid reaction of SiO (g) and C(s). It is more likely that the whisker/fiber formed via gas-gas reaction between SiO (g) and CO (g) such that the morphology of the SiC could be totally independent of the morphology of carbon sources. The reaction can proceed as follows: [5]

$$SiO_{(g)} + 3CO_{(g)} \longrightarrow SiC_{(s)} + 2CO_{2\,(g)} \quad \text{..................6}$$

Reaction (6) favors the growth of SiC whiskers that are similar to those obtained by chemical vapor deposition with SiO and CO as the primary reactants in the literature. [41, 42]
In powder route, reaction of gaseous SiO into Si and SiO_2 can take place according to the reaction[43, 44].

$$2SiO_{(g)} \longrightarrow Si_{(s)} + SiO_2 \quad \text{..................7}$$

The resulting Si would react with carbon to give rise to spherical SiC particles, according to the following reaction.

$$Si_{(s)} + C_{(s)} \longrightarrow SiC_{(s)} \quad \text{..................8}$$

Therefore, in solution route gas–gas reactions and in powder route gas–solid reaction occurred during the preparation of porous SiC ceramic and this is supported by SEM analysis.

3.5 Surface Morphology

The surface morphology of silicon carbide prepared by powder and solution route was studied using SEM. The micrographs are given in figure 5.

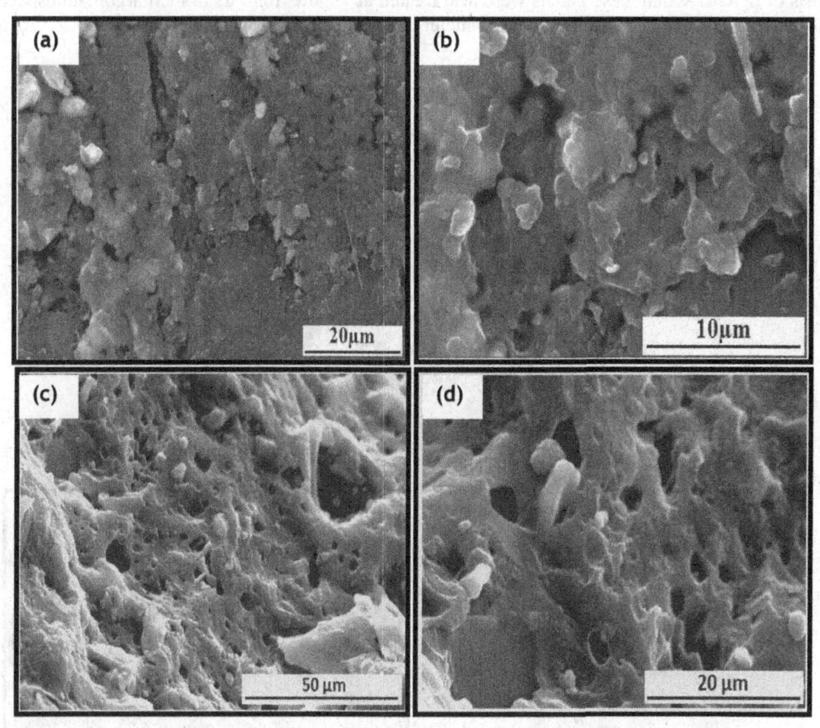

Figure 5. Scanning electron micrograph (a), (b) Green pellet obtained from powder route and (c), (d) Green pellet obtained from solution route

Fig.5 shows SEM micrograph of green pellets obtained by powder and solution route. These pellets have porous structure (fig.5[c & d]). Green pellets made from powder route were less porous in nature (fig.5 a & b). The micrographs of green pellets (made from solution route) after heat treatment to 16500 C in inert atmosphere are shown in figure 6. It shows SiC, whiskers formation (fig.6c) and the diameter of whiskers ranged from 0.12μm to 0.17μm and length of whiskers ranged formed 9.5mμ to 11.6μm. The obtained SiC ceramic was very fluffy (fig.6) in nature.

SEM micrograph of green pellets obtained from powder route shows that wood and sílica particles were coated with phenolic resin Fig 5 (a) & (b). Also the pore formation was less during synthesis of pellets. When these pellets were heat treated at 1650^0C for 4hrs in inert argon atmosphere pore formation occurred. The average pore diameter of obtained bio SiC is ranged in between 2.5μm to 6.75μm. (fig.7)

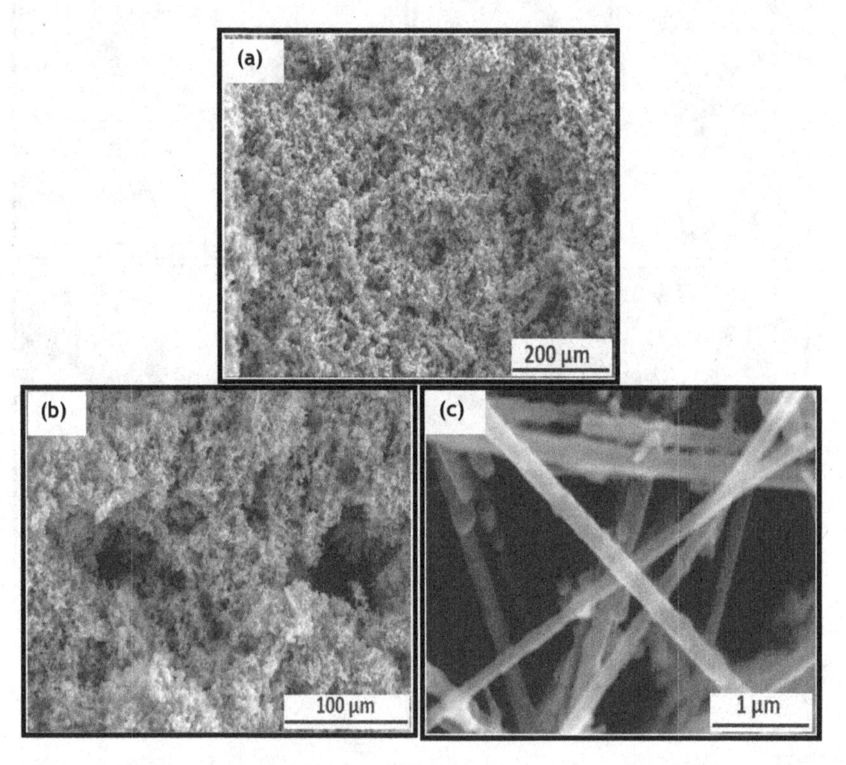

Figure 6. Scanning electron micrograph of (a), (b) SiC ceramic and (c) SiC whiskers

[Obtained from solution route]

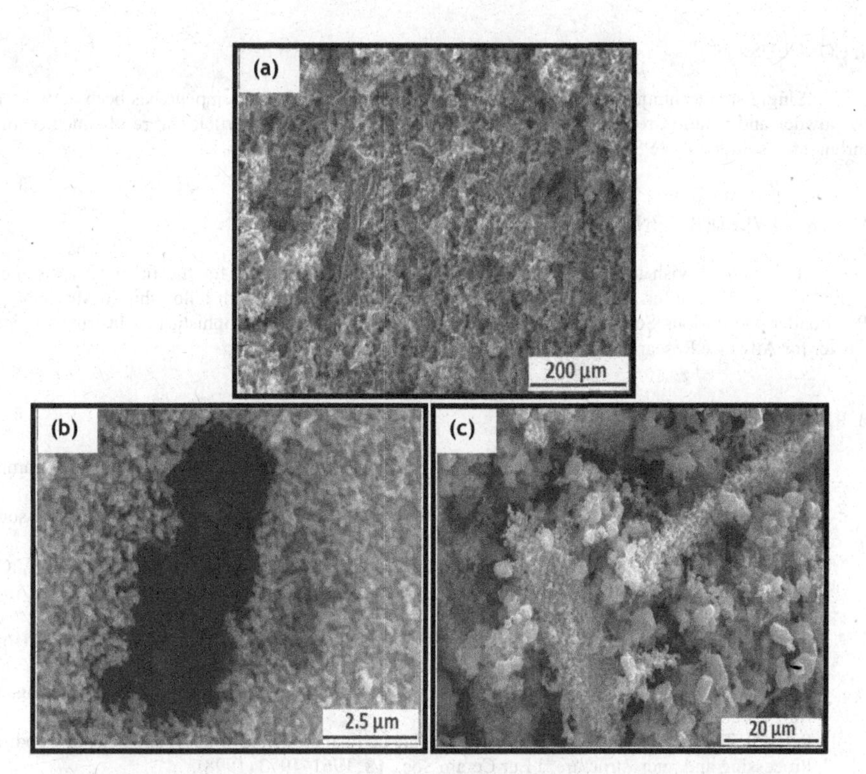

Figure 7. Scanning electron micrograph of (a), (b) SiC ceramic and (c) Particles of SiC

[Obtained from powder route]

4. CONCLUSION

Single-step technique for the synthesis of porous SiC using wood template has been developed by powder and solution route. In the powder form, the spherical SiC particles were obtained, while through the solution route SiC whiskers are formed.

5. ACKNOWLEDGEMENT

The authors wish to thank University Grants Commission, India for the financial assistance provided under Centre for Advanced Studies Program of UGC and research fellowship to Mr.Hemang Patel under Meritorious Scheme of UGC. The XRD work was done at Sophisticated Instrumentation Center for Advance Research and Testing (SICART), Vallabh Vidyanagar

6. REFERENCES

[1] P. Colombo, and E. Bernardo, Macro- and Micro-cellular Porous Ceramics from Preceramic Polymers, Composites Sci. Tech., **63**, 2353-2359 (2003).

[2] P. Ciambelli, V. Palma, P. Russo, and S. Vaccaro, Performances of a catalytic foam trap for soot abatement , Catalysis Today., **75**,471-478 (2002).

[3] M.D.M. Innocentini, A.R.F. Pardo, B.A. Menegazzo, L.R.M. Bittencourt, R.P. Rettore, V.C. Pandolfelli, Permeability of High-alumina Refractory on Various Hydraulic Binders, J. Am. Ceram.Soc., **85**, 1517-1521 (2002).

[4] P. Sepulveda, Gelcasting Foams for Porous Ceramics,Am. Ceram. Soc. Bull., **76[10]**, 61-65 (1997).

[5] L. Montanaro, Y. Jorand, G. Fantozzi, and A. Negro, Ceramic foams by powder processing, J. Eur. Ceram.Soc., **18**, 1339-1350 (1998).

[6] P. Greil, T. Lifka, and A. Kaindl, (a) Biomorphic cellular silicon carbide ceramics from wood: I. Processing and microstructure, J Eur Ceram Soc., **18**, 1961–1973 (1998).
(b) P. Greil, T. Lifka, and A. Kaindl, (b) Biomorphic cellular silicon carbide ceramics from wood: II. Mechanical properties, J Eur Ceram Soc., **18**, 1975–1983 (1998).

[7] K. Schwartzwalder, and A.V. Somers, Method of making porous ceramics articles, (US Pat. No. 3 090 094, May 21, (1963).

[8] J. Saggio-Woyansky, and C.E. Scottetal, Processing of Porous Ceramics, Amer. Ceram. Soc. Bull., **71**, 1674-1682 (1992).

[9] E. Sousa, C.B. Silveira, T. Fey, P. Greil, D. Hotza, and A.P.N.Oliveira, LZSA glass ceramic foams prepared by replication process, Adv. Appl. Ceram., **104**, 22-29 (2005).

[10] P. Sepulveda, and J.G.P. Binner, Processing of Cellular Ceramics by Foaming and in situ Polymerisation of Organic Monomers, J. Eur. Ceram. Soc., **19**, 2059-2066 (1999).

[11] H.X. Peng, Z. Fan, J.R.G. Evans, and J.J.C. Busfield, Microstructure of ceramic foams,J. Eur. Ceram.Soc., **20**, 807-813 (2000).

[12] A.H. Heuer, D.J. Fink, V.J. Arias, P.D. Calvert, K. Kendali,G.L. Messing, J. Blackwell, P.C. Rieke, D.H. Thompson,A.P. Wheeler, A. Veis, and A.I. Caplan Innovative Materials Processing Strategies: A Biomimetic Approach, web of Science., **255[5048]**, 1098-105 (1992).

[13] C.E. Byrne, and D.E. Nagle, Cellulose-derived composites—a new method for materials processing, Mat. Res. Innovat., **1**, 137-145 (1997).

[14] H. Sieber, Biomimetic synthesis of ceramics and ceramic composites, Mat. Sci Eng A., **412**, 43-47 (2005).

[15] E. Vogli, H. Sieber, and P. Greil, Biomorphic SiC-ceramic prepared by Si-vapor phase infiltration of wood, J Eur Ceram Soc., **22**,2663–2668 (2002).

[16] C.R. Rambo, J. Cao, O. Rusina, and H. Sieber, Manufacturing of biomorphic (Si, Ti, Zr)-carbide ceramics by sol–gel processing, Carbon., **43**,1174–1183 (2005).

[17] D.A. Streitwieser, N. Popovska, H. Gerhard, and G. Emig, Application of the chemical vapor infiltration and reaction (CVI-R) technique for the preparation of highly porous biomorphic SiC ceramics derived from paper, J Eur Ceram Soc., **25**, 817–828 (2005).

[18] O. Chakrabarti, L Weisensel, and H.Sieber, Reactive melt infiltration processing of biomorphic Si–Mo–Cceramics from wood, J Am Ceram Soc., **88(7)**,1792–1798 (2005).

[19] J. Martínez-Fernández, F.M. Valera-Feria, and M.Singh, High temperature compressive mechanical behavior of biomorphic silicon carbide ceramics, Sci Mater., **43**,813–818 (2000).

[20] A.R. de Arellano-López, J. Martínez-Fernández, F.M. Varela-Feria, T.S. Orlova, K.C. Goretta, F. Gutierrez-Mora, N. Chen, and J.L. Routbort, Erosion and strength degradation of biomorphic SiC, J. Eur. Ceram Soc.,**24**,861–870 (2004).

[21] F. Gutierrez-Mora, K.C. Goretta, F.M. Varela-Feria, A.R.Arellano López, and J .Martínez Fernández Indentation hardness of biomorphic SiC, Int J Refract Metab Hard Mater., **23**,369–374 (2005).

[22] M .Presas, J.Y. Pastor, J. Llorca, A.R. Arellano López, J. Martínez Fernández, R. Sepúlveda, Microstructure and fracture properties of biomorphic SiC, Int J. Refract Metab Hard Mater., **24**,49–54 (2006).

[23] H-T. Fang, Z-D .Yin, J-C Zhu, J-H Jeon, and Y-D Hahn, Effect of Al additive in Si slurry coating on liquid Si infiltration into carbon–carbon composites, Carbon., **39**,2035–2041 (2001).

[24] G. Hou, Z. Jin, and J. Qian, Effect of holding time on the basic properties of biomorphic SiC ceramic derived from beech wood, Mater Sci Eng A., **452**,278–283 (2007).

[25] M. Singh, and Y. Bo-Moon, Reactive processing of environmentally conscious, biomorphic ceramics from natural wood precursors, J. Eur. Ceram. Soc., **24**,209–217 (2004).

[26] M. Singh, and J.A. Salem, Mechanical properties and microstructure of biomorphic silicon carbide ceramics fabricated from wood precursors, J.Eur. Ceram. Soc., **22**, 2709–2717 (2002).

[27] P. Colombo, Conventional and Novel Processing Methods for Cellular Ceramics, Philos. Trans. Roy. Soc. A., **364**, 109–24 (2006).

[28] M. Scheffler, and P. Colombo, Cellular Ceramics: Structure, Manufacturing, Properties and Applications, Weinheim Wiley-VCH., Germany (ISBN 3-527-31320-6), p. 645 (2005).

[29] D. J. Green and P. Colombo, Cellular Ceramics: Intriguing Structures, Novel Properties, and Innovative Applications, Mater. Res. Soc. Bull., **28 [4]**, 296–300 (2003).

[30] T.X. Fan, T. Hirose, T. Okabe, D. Zhang, R. Teranisi, and M.Yoshimura, effect of components upon the surface area of wood ceramics, J. Porous Mater., **9**, 35-42 (2002).

[31] P. Greil, Biomorphous ceramics from lignocellulosics, J. Eur. Ceram. Soc., **21**, 105-118 (2001).

[32] S. Chand, Carbon fibers for composites, *J Mater Sci*.,**35**, 1303–13 (2000).

[33] Y. Yamashita, K. Ouchi, A study on carbonization of phenol-formaldehyde resin labelled with deuterium and ^{13}C ,Carbon., **19**, 89-94 (1981).

[34] T.Hirose, T.X.Fan, T.Okabe, and M.Yoshimura, Effect of carbonizing speed on the property changes of woodceramics impregnated with liquefacient woods, Mater.Lett.,**52**p 229-233(2002).

[35] K. S. W.Sing, D. H.Everett, R. A. W. Haul, L. Moscou, R. A. Pierotti, J. Rouquerol, T. Siemieniewska, reporting physisorption data for gas/solid systems with Special Reference to the Determination of Surface Area and Porosity. Pure & Appl. Chem.., **57(4)**, 603-619 (1985).

[36] L.K.Frevel, D.R.Petersen, and C.K.Saha, Polytype distribution in silicon carbide, J.Mater.Sci., **27**, 1913-1915 (1992).

[37] I. Hasegawa, T. Nakamura, S. Motojima, and M. Kajiwara, Silica gel-phenolic resin hybrid fibres: new precursors for continuous beta-silicon carbide fibres. J. Mat. Chem., **5(1)**, 193–194 (1995).

[38] X. K. Li, L.Liu, Y. X. Zhang, S. D. Shen, S. Ge, and L. C.Ling, Synthesis of nanometre silicon carbide whiskers from binary carbonaceous silica aerogels, Carbon., **39(2)**, 159–165 (2001).

[39] C.Vix-Guterl, and P. Ehrburger, Effect of the properties of a carbon substrate on its reaction with silica for silicon carbide formation, Carbon., **35(10–11)**, 1587–1592 (1997).

[40] C. Vix-Guterl, I.Alix, P.Gibot, and P.Ehrburger, Formation of tubular silicon carbide from a carbon-silica material by using a reactive replica technique: infra-red characterization, Appl. Surf. Sci., **210(3–4)**, 329–337 (2003).

[41] C.Vix-Guterl, B. McEnaney, and P. Ehrburger, SiC material produced by carbothermal reduction of a freeze gel silica-carbon artifact, J. Eur. Ceram. Soc., **19(4)**, 427–432 (1999).

[42] M. Saito, S. Nagashima, and A. Kato, Crystal growth of SiC whisker from the $SiO_{(g)}$–CO system, J. Mat. Sci. Lett., **11(7)**, 373–376 (1992).

[43] V. D. Krstic, Production of fine, high-purity beta silicon carbide powders, J. Am. Ceram. Soc., **75(1)**, 170–174 (1992).

[44] W. S. Seo, and K. Koumoto, Stacking faults in b-SiC formed during carbothermal reduction of SiO_2, J. Am. Ceram. Soc., **79(7)**, 1777–1782 (1996).

FABRICATION OF BETA-CRISTOBALITE POROUS MATERIAL FROM DIATOMITE WITH SOME IMPURITIES

Osman Şan*, Cem Özgür and Remzi Gören
Dumlupinar University, Department of Ceramic Engineering, Kütahya, Turkey

ABSTRACT

The room temperature stabilized β-cristobalite ceramic has great potential to use in production of engineering ceramic materials due to its high resistance to thermal shock and low expansion coefficient with high chemical resistance and low density. The material was investigated from the mixture of purified diatomite doped with alumina and calcium ions obtained from nitrates. The diatomite was low grade material (68.08 wt.% SiO_2) and its purification was achieved by hot acid leaching for 72 h and obtained relatively high grade-silica powder (96.41 wt.% SiO_2). The powder was uniaxially pressed at 15 kPa, and later sintered at 1000°C for 24 h and at 1300°C for one minute. The material sintered at low temperature has promising engineering properties: the thermal expansion is almost linearly and found the thermal expansion coefficient as 11.8×10^{-6} °C^{-1}. High temperature sintering leads to cristobalite crystallization and thus the thermal stability of the product sample was significantly decreased.

1. INTRODUCTION

The diatomite raw materials are high abundance in nature and mostly contain alumina with the other impurities such as alkaline earth, alkaline metals, and iron with and without organic components. After some purification, the material can be used extensively as fillers, filter aids, abrasives, insulating materials, conventional catalyst supports, membranes and biocatalytic proteins and cells. The beneficiation processes comprise crushing, drying, calcinations, classification and furthermore hydrothermal treatments (leaching). After then the materials powdered using attrition milling or planetary. By this way, the particles having micron to nano sizes with less impurity could be obtained.

High-grade diatomaceous earth containing a minimum of about 95% diatomite ($SiO_2 \cdot nH_2O$) is less abundant in nature and mostly contains impurities. Generally, the crystalline nature of diatomite materials was of X-ray amorphous. Because of the impurities, the sintering of diatomite's at high temperature leads to crystallization. The crystallizations are not controlled process where the materials generally appeared as α-cristobalite or cristobalite like structures. To achieve of the crystallization as β-cristobalite makes the material potential to use widely for industrial applications. Low and linearly thermal expansion behaviour, besides other properties such as chemical inertness and less densities are obtained. The low thermal expansion behaviour with linear changes minimises the occurrence of thermal stresses on ceramic material during abrupt temperature change [1]. These useful properties make the material a viable alternative too many conventional ceramics used in harsh thermal environments.

* Corresponding author: Osman Şan, Dumlupinar University, Department of Ceramic Engineering, Kütahya, 43100, Turkey. Tel: +90 274 265 20 31/4302, Fax: +90 274 265 20 66, E-mail: osmansan@dumlupinar.edu.tr.

The β-cristobalite ceramic powder could be obtained at room temperature by phase stabilization using "stuffing" cations. The incorporation of foreign ions (Ca, Sr, Cu, Na) in the interstices of silicate structure is charge-compensated by the substitution of Al^{3+} for Si^{4+} in the framework. Recent studies indicated that the purification of diatomite by hot-acid leaching 2 and stabilization of the powder as β-cristobalite 3 is possible. The powder was used for fabrication of relatively high porosity material with thermally stable. The success can be explained by the purification of the sample and later stabilization using aluminum and calcium ions. The purification of diatomite was achieved by acid leaching and obtained a high silica-containing material (98 wt.% SiO_2). The impurities are Al_2O_3, CaO, K_2O, MgO, Fe_2O_3, and TiO_2. The material designed as β-cristobalite using the composition $Si_{1-x}Al_xCa_{x/2}O_2$ where the best being x = 0.05. The required stuffing ions were determined by considering the impurities (Al and Ca) of the diatomite. The others were neglected. When the neglected impurities attended, the formula of β-cristobalite changed as: $Si_{0.937}Al_{0.049}Ca_{0.024}Na_{0.005}K_{0.004}Mg_{0.008}Fe_{0.001}Ti_{0.003}O_2$. The impurities have no significant influence on the thermal behaviour of the ceramic material for sintering applied at 1000°C where the thermal expansion coefficient was found as $11.42\times10{-6}$ °C^{-1}. These useful properties make the material a viable alternative too many conventional ceramics used in harsh thermal environments.

The above materials has been obtained from diatomite by long periods leaching such that 240 h. In this study, the β-cristobalite material was investigated from the same diatomite achieving through less time leaching where the process was applied for 72 h. In this case, the material has more impurities but the cost of powder processing was reduced. It enhanced the use of low-grade diatomite materials in fabrication of high-performance ceramic materials for use in thermal environments.

2. MATERIALS AND METHODS

Beta-cristobalite ceramic was investigated from diatomite powder after hot acid leaching. The material was stabilized with Al^{3+} and Ca^{2+} as stuffing ions. After the stabilized β-cristobalite samples had been produced, the phases were identified by X-ray diffraction (XRD) analysis and characterised according to their thermal expansion behaviour. SEM images of the materials were also taken and studied.

The starting powder of β-cristobalite ceramic was a raw diatomite obtained from the Kütahya region of Turkey. The main chemical composition was as follows (wt.%): 68.08 SiO_2, 17.99 Al_2O_3, 4.22 MgO, 3.36 Fe_2O_3, 1.32 K_2O, 0.98 CaO, 0.67 Na_2O and 0.29 TiO_2. The sample was ground in an attrition mill for 1 h using alumina balls with an aqueous system. The slurry was dried in 105°C for 24 h, and then sieved through an aperture of 45 m and stored for the leaching. For the leaching experiments, the prepared diatomite powder was treated in HCl solutions at 75°C. The solutions contained 5 M HCl acid and the treatment time by a magnetic stirrer was applied as 72 h. During leaching, 10 g of diatomite sample was weighed and poured in 200 ml solution, then the solution was stirred continuously at 500 rpm and a thermostat was employed to keep the reaction medium at constant temperature. After the leaching operation, the solid product was filtered and washed with distilled water.

The β-cristobalite material was fabricated from the mixture of the purified diatomite powder and metal nitrates. The nitrates were obtained from Merck: $Al(NO_3)_3{\cdot}9H_2O$ (CAS No.: 7784-27-2) and $Ca(NO_3)_2{\cdot}4H_2O$ (CAS No.: 13477-34-4). The β-cristobalite composition was designed as $Si_{0.95}Al_{0.05}Ca_{0.025}O_2$ where some impurity from the purified-diatomite was neglected. The nitrates were dissolved within water; the purified diatomite powder was then added and dried to obtain a powder mixture. The powder mixture was agglomerated, uniaxally pressed at 15 kPa, and later sintered at

1000°C for 24 h and at 1300°C for one minute. The heating and cooling rate of the furnace was 5°C/min.

The thermal expansion behaviour of the diatomite-based materials was compared with a fully β-cristobalite sample. The reference sample was prepared from colloidal silica and the stuffing ions are the same as for the above diatomite production. The sintering temperature of the reference sample was of 1000°C for 24 h.

The study of the samples included: (i) chemical composition measurement by X-ray fluorescence (Spectro X-LAB 200), (ii) crystalline phase identification by X-ray analysis (Rigaku Miniflex powder diffractometer employing CuKα radiation in 2θ = 10-65° at a ganiometer rate of 2θ = 2°/min.), (iii) thermal expansion behaviour determined by dilatometry (Netzch DIL 402 PC) using 2.5 cm long rods through the samples heated at the rate of 10°C/min, and (iv) microstructural analysis using a SEM (Zeiss Suprat 50).

3. RESULTS AND DISCUSSIONS

The chemical composition of leached diatomite was as follows (wt.%): 96.41 SiO_2, 1.15 Al_2O_3, 0.12 Na_2O, 0.45 K_2O, 0.19, 0.36 MgO, 0.10 Fe_2O_3, 0.23 TiO_2. In here, the purification achieved for leaching times as 72 h and obtained relatively high silica-containing material where the raw diatomite is a low grade material (68.08 wt.% SiO_2). The X-ray analysis of the raw and leached-diatomite powder is given in Fig. 1. It indicates a significant amount of crystalline phases such as quartz, aluminum silicates, magnesium silicates and plagioclases. The crystalline phases of the sample undergo considerable change with the leaching and the silicates, especially Mg silicates, except quartz and plagioclase, disappear.

The leached-powder was agglomerated, uniaxally pressed and sintered at 1000°C for 24 h and at 1300°C for one minute. The crystalline phases and microstructure of the samples were investigated. At low-temperature sintering, the crystallisation is well formed and shows β-cristobalite with some crystalline impurities (see Fig. 2). In that case, a significant amount of quartz phase is determined and the minor constituent is of plagioclase. The plagioclase crystallisation is not desirable if the material used for fabrication is of high porosity in which the flux impurity fuses the particles and closes the micropores of the diatomites. The high temperature sintering decreased the amount of quartz crystallisation, but this time the high temperature state produced α-cristobalite crystallisation. Alpha-cristobalite crystallisation is totally undesirable for the material used in thermal environments in which nonlinear thermal expansion occurs during the α-form displacive phase transition.

Beta-cristobalite reference material was prepared for produce of a comparative study. In this case, the parent material was colloidal silica with stuffing ions such as Al^{+3} and Ca^{+2} obtained from the same nitrates as used in the diatomite stabilisation. The X-ray analysis clearly indicates that the reference β-cristobalite material could be obtained without any phase impurity (see Fig. 3).

The microstructure of diatomite particles after purification and its sinter ceramic components as such was investigated by the scanning electron microscope (SEM). Naturally, some impurities were deposited on the raw diatomite particles [2,4]. The leaching operation removed the impurities and thus the typical diatomite type microscopic structure could be obtained. Fig. 4 shows the SEM micrograph of a typical facture edge of the ceramic materials obtained from the purified and stabilized diatomite powder. The sintering temperatures were of 1000°C for 24 h (a) and 1300°C for 1 min (b). The micrographs show the better microstructural features in which the homogenous pore distributions could be obtained. The amount of apparent porosity of the samples were 50.45% and %46.29% through the materials sintered at 1000°C for 24 hours and 1300°C for one minute, respectively. This type of microstructure is believed to be a great potential to use these materials as ceramic filters or substrate for different applications.

The diatomite without purification (raw diatomite) was shaped and sintered at 1300°C to confirm the statements given above and found that the microstructure is non-porous (see Fig. 5).

In our study, a fully beta-cristobalite material was fabricated and its thermal behaviour was determined and is given in Figure 6a. Poor thermal stability of α-cristobalite material is well-known; sudden increase of thermal expansion occurs at temperatures between 170 and 270°C. The ceramic material obtained from diatomite powder without stabilization has almost fully α-cristobalite crystallization 3 . The thermal expansion behaviour of α-crystallized material has been shown in Perrota's studies [5] and is presented in Fig. 6b. The large difference in thermal stabilities between the alpha- and beta-crystallisation is clearly shown. The same figure also shows the thermal behaviours of presently fabricated ceramic materials in which the sintering temperatures were of 1000°C for 24 h (c) and 1300°C for 1min (d). The material sintered at low temperature has promising engineering properties: the thermal expansion is almost linearly and found the thermal expansion coefficient as 11.8×10^{-6} $°C^{-1}$. The material sintered at high temperature is quite different: it has a small quantity of α-cristobalite crystallization (see Fig. 2). This study also emphasize that a small quantity of α-phase in the composition adversely influenced the thermal stability of the product samples.

4. CONCLUSIONS

This study indicated that the possibility of fabrication thermally stable ceramic material from a low grade diatomite. The success can be explained by the sufficiently purification of the sample and later stabilization using aluminum and calcium ions. The purification of diatomite was achieved by acid leaching and obtained a high silica-containing material (96.41 wt.% SiO_2). The impurities are Al_2O_3, CaO, K_2O, MgO, Fe_2O_3, and TiO_2. The thermally stable ceramic material designed as β-cristobalite using the composition $Si_{0.95}Al_{0.05}Ca_{0.025}O_2$. The required stuffing ions were determined by considering the impurities (Al and Ca) of the diatomite. The others were neglected. When the neglected impurities attended, the formula of β-cristobalite changed as: $Si_{0.937}Al_{0.049}Ca_{0.024}Na_{0.005}K_{0.004}Mg_{0.008}Fe_{0.001}Ti_{0.003}O_2$. But, the impurities have no significant influence on the thermal behaviour of the ceramic material for the material sintered at 1000°C where the thermal expansion coefficient was found as $11.8\times10{-}6$ $°C^{-1}$. The thermal expansion coefficient of the material obtained with long time leaching (240 h) had been $11.42\times10{-}6$ $°C^{-1}$ [3]. No significant difference with the present study indicated that the leaching time at 72 h is good enough for the diatomite purification.

REFERENCES

[1] C.H. Chao, H.Y. Lu, Stress-induced β→α-cristobalite phase transformation in (Na_2O+Al_2O_3)-codoped silica, Materials Science and Engineering A., 328 (2002) 267–276.

[2] O. Şan, R. Gören, and C. Özgür, Purification of diatomite powder by acid leaching for use in fabrication of porous ceramics, Internal Journal of Mineral Processing, 93 (2009) 6-10.

[3] O. Şan, and C. Özgür, Preparation of a stabilized cristobalite ceramic from diatomite, Journal of Alloys and Compounds, 484 (2009) 920-923.

[4] H. Hadjar, B. Hamdi, M. Jaber, J. Brendlé, Z. Kessaïssia, H. Balard, J.B. Donnet, Elaboration and characterisation of new mesoporous materials from diatomite and charcoal, Microporous and Mesoporous Materials, 107 (2008) 219–226.

[5] A.J. Perrota, D.K. Grubbs, E.S. Martin, H.R. Dando, H.A. McKinstry, C. Huang, Chemically stabilization of β-cristobalite, Journal of American Ceramic Society, 72 (1989) 441447.

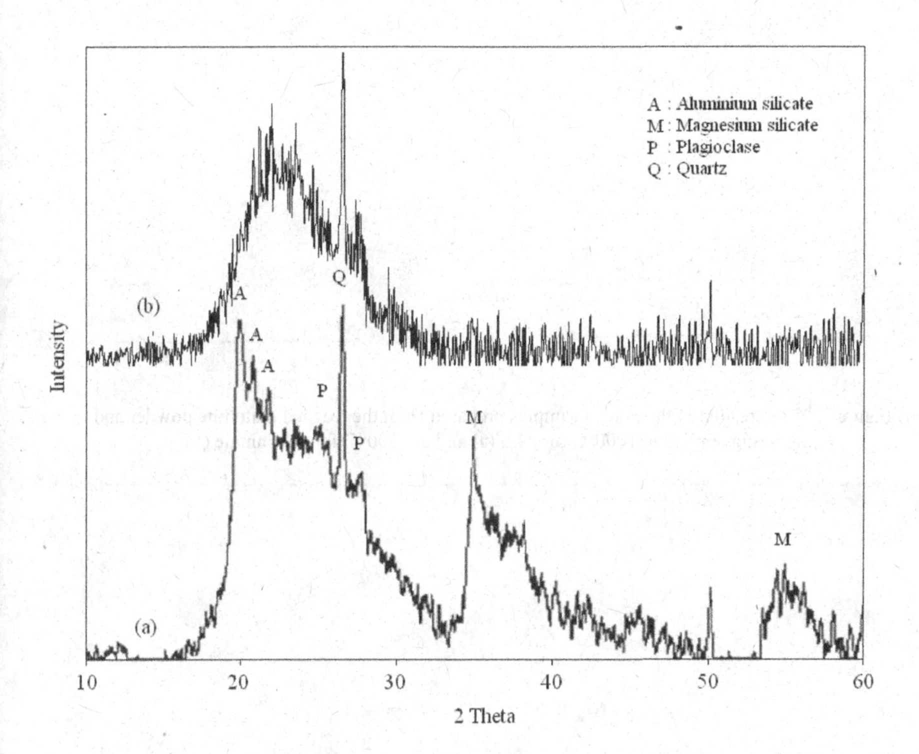

Figure 1. Phase compositions of the as received (a) and leached diatomite (b).

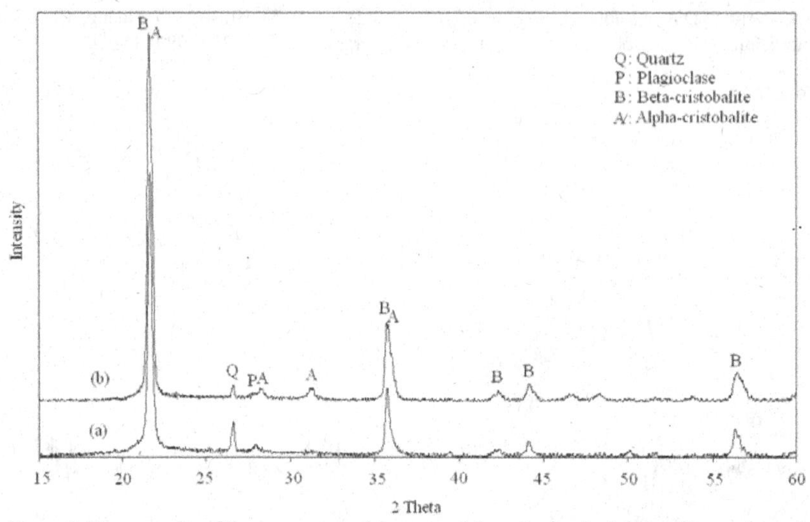

Figure 2. X-ray results of the porous samples prepared from the leached diatomite powder and sintering applied at 1000°C for 24 h (a) and at 1300°C for one minute (b).

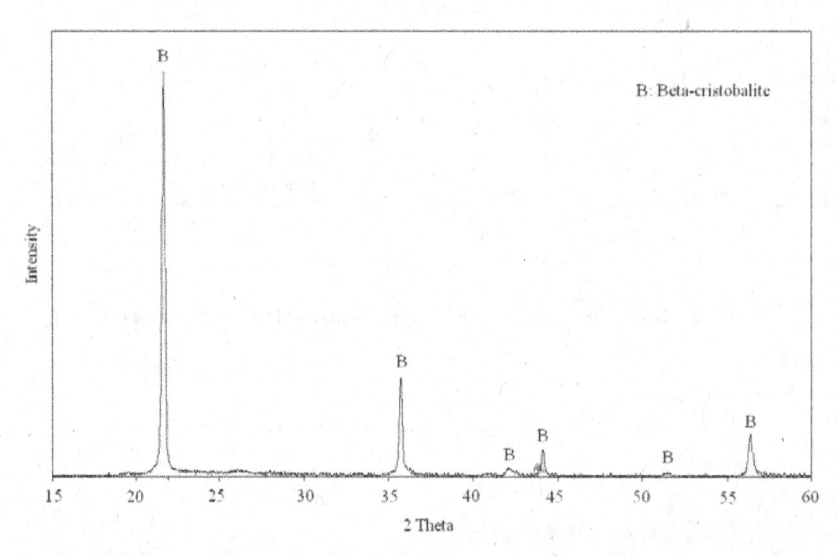

Figure 3. X-ray results of the beta-cristobalite sample prepared from colloidal silica.

4b

Figure 4. SEM photograph of the porous materials produced from the leached diatomite with sintering at 1000°C for 12 h (a) and sintering at 1300°C for one minute (b).

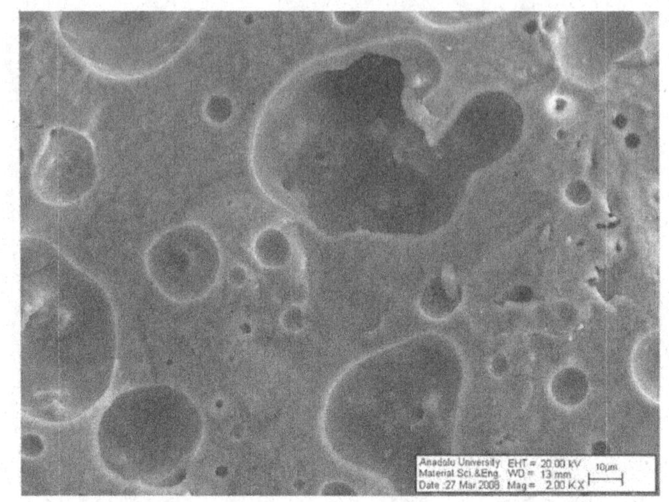

Figure 5. SEM picture of ceramic material produced from the diatomite without leaching.

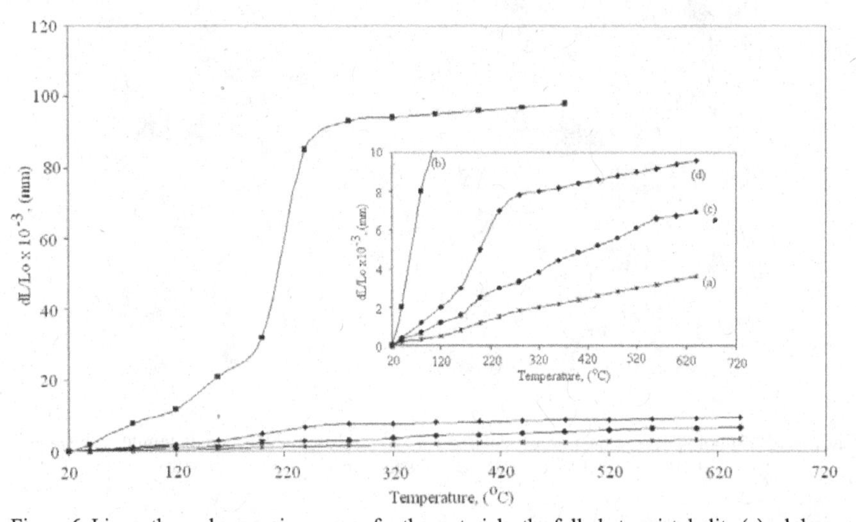

Figure 6. Linear thermal expansion curves for the materials: the fully beta-cristobalite (a), alpha-cristobalite (b) 3 , sintering at 1000°C for 12 h (c) and sintering at 1300°C for one minute (d).

MICROSTRUCTURAL STUDY OF ALUMINA POROUS CERAMIC PRODUCED BY REACTION BONDING OF ALUMINIUM POWDER MIXED WITH CORN STARCH

Juliana Anggono*, Ida A.O.R.S. Shavitri, and Soejono Tjitro
Mechanical Engineering Department
Petra Christian University, Surabaya 60236, Indonesia

ABSTRACT

Addition of corn starch to the Al powders in a reaction bonding process has resulted in alumina porous ceramics. The corn starch added was meant to create porosities during heating therefore they act as channels for oxygen to access the inner part of the samples for oxidation. The number of corn starch added to the Al powder was varied from 10%, 20%, and 30 wt. %. The green bodies of pellet samples with diameter of 20 mm and 4 mm height were heated at various temperatures of 1000°C, 1200°C, and 1400°C with a heating rate of 1.5C/minute in air. Microstructural study was performed using scanning electron microscope (SEM) and supported by phase identification using XRD and EDAX. XRD and EDAX have identified the presence of α-Al_2O_3, Fe, and Si which are impurities of Al powders. From SEM, it shows that the outer surface of all samples is denser than the inner part and identified as Al_2O_3. The thickness of this dense outer skin increases with an increase in heating temperature and the number of corn starch added. After sintering to 1400°C, smaller size of pores (0.005 μm) originated from the interparticles stack while the bigger pores (0.008 μm and 0.0175 μm) obtained from the sites left by the corn starch particles were identified by porosimeter test.

1. INTRODUCTION

Several processing routes are nowadays available for the production of porous ceramics. A review on processing routes to macroporous ceramics has been published by Studart, *et al.* which they did comparison in some of the processing routes currently available for the preparation of macroporous ceramics (i.e. pore size > 50 nm).[1] Porous ceramics have been widely produced for molten metal filter, refractory insulation, water purification catalyst support, electrodes and support for batteries and solid oxide fuel cells, scaffolds for bone replacement and tissue engineering, and heating elements. Their applications where high temperatures, high wear rate, and corrosive environment are involved need materials that have high melting point, excellent corrosion and wear resistance together with low thermal conductivity, controlled permeability, low density, high specific strength, and low dielectric constant. The properties of porous ceramics can be tailored by controlling and optimising the composition and microstructure of porous ceramics, i.e. pore volume fraction, pore structures, and grain bonding.

Porous Al_2O_3 ceramics have attracted considerable attention for decades because of their good thermal stability at elevated temperatures. Both conventional [2-6] and novel methods [7-10] have been used

* Corresponding author, member of American Ceramic Society
Mechanical Engineering Department
Petra Christian University
Jalan Siwalankerto 121-131 Surabaya 60236
Indonesia
Email: julianaa@petra.ac.id, telp: +62-31-78009138, fax: +62-31-849 1214

and developed to fabricate porous Al_2O_3 ceramics for structural and bioceramics applications. They differ in terms of processing techniques and final microstructures achieved.

In the present study, microstructural development of reaction-bonded aluminium oxide (RBAO) which prepared from a mixture of Al flakes powders and corn strach was investigated. Corn starch was used to enhance the oxidation kinetics of Al powders through pores formation at 1400°C. Al flakes powder has the highest degree of oxidation compared to other morphology, i.e. irregular and spherical.[11]

2. EXPERIMENTAL PROCEDURE

The Al flakes powder used (GLORIA type 2377) is shown in Figure 1a. They were plate like with a thickness < 1 μm and diameters < 80 μm and had a high level of oxidation. The chemical composition of the Al flakes supplied by the manufacturer was 90% with the only significant metallic impurity was Fe. The corn starch (HONIG) used as the pore former and it can be found in the general food store for cooking. The starch powder varies in morphology and size as shown in Figure 1b.

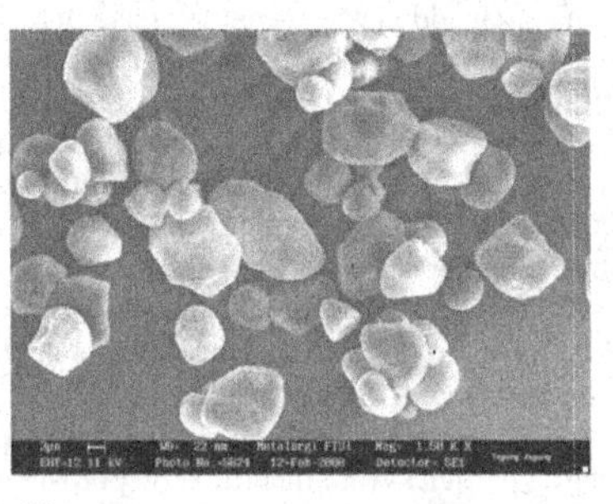

a) b)
Figure 1 SEM photographs of a) Al flakes powder and b) corn starch used as pore former.

Samples were prepared by adding 10, 20, 30 wt. % of corn starch to the Al powder and then mixed them using ball milling at 64 rpm for 8 hours. Table 1 shows the relative amounts of Al and corn starch in the mixture and the notation used for samples' identification.

The homogeneous mixture obtained was shaped in a steel die press at 10.34 MPa to form green compacts in pellet form with Ø= 20 mm and h= 4 mm. Green compacts were sintered in air atmosphere at temperatures of 1000, 1200, and 1400°C in a muffle furnace (Naber, type N11-220) and in a tube furnace. The sintering process was done in several heating steps as follows: heating to 300°C at 1.5° C $minute^{-1}$, followed by an isothermal hold for 1 hour; heating to 600°C at 1.5° C $minute^{-1}$ followed by an hour isothermal hold to burn out the starch, another isothermal hold for 1 hour at 900°C and 1000°C to convert the Al to the corresponding oxide. Sintering at temperatures of 1200°C and 1400°C followed similar steps as heating up to 1000°C which then followed by a final heating ramp to 1200° C and 1400°C at 1.5° C $minute^{-1}$ with an hour isothermal hold to complete the oxidation of Al. Samples were then cooled in the furnace. Final microstructures obtained after heating the Al/corn starch mixtures were investigated by XRD and SEM/EDAX (jsm-63601a, JEOL, Tokyo, Japan).

Table 1 Relative Amounts of Al and Corn Starch Mixed in Different Ratio of Corn Starch/Al Flakes and Notation Used for Samples' Identification.

Component	Weight (g)		
	Composition A (10 wt.% corn starch)	Composition B (20 wt.% corn starch)	Composition C (30 wt.% corn starch)
Al	1.8	1.6	1.4
Corn starch	0.2	0.4	0.6

Numbers 1, 2, 3 after each letter (A, B, C) are used to identify sintering temperature of 1000°C, 1200°C, and 1400°C respectively, e.g. B1 shows identification of samples with composition B after sintering at 1000°C.

3. RESULTS AND DISCUSSION

Phase Evolution During Heat Treatment

The structural evolution of all heat treated samples was monitored by XRD. The effect of corn starch addition to the Al powder on phase evolution is discussed. Figures 2-4 show a series of XRD patterns obtained from all samples after sintering at different temperature (1000°C to 1400°C). Some of these samples were also used for microstructural observation, in order to obtain more information on the reaction sequence during heating of the corn starch/Al blends. All samples were examined at room temperature after reaching the indicated temperature during the heat treatment.

The reaction products after heating at 1000°C are mainly α-Al_2O_3 with a small quantity of -Al_2O_3 as transition phase detected at this temperature. This might be because of the Al particle size used are relatively large and the aluminium oxide would only formed at quite high temperature at which the α-Al_2O_3 is stable.

The TGA/DSC study on Al powder/polymethylsiloxane blends done by Anggono and Derby indicated that significant oxidation of Al started at a temperature >600°C and oxidation is complete by 1400°C.[10-11]

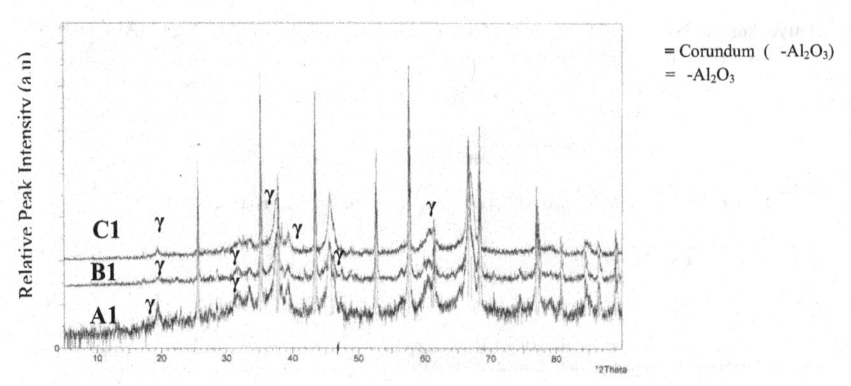

Figure 2 XRD traces taken from A1, B1, and C1 samples after sintering at 1000°C.

XRD data from samples A1 heated to 1000°C (Fig. 2) shows the -Al_2O_3 peaks are a little more distinct compared to samples B1 and C1. As the temperature increased to 1200°C, -Al_2O_3 phase was still found in the structure of samples A2 and B2 (Figure 3). Phase of -Al_2O_3 may also exist in sample C2 but with low intensity which can be below the detection limit for XRD. Iron oxide (Fe_3O_4) and sodium magnesium aluminium oxide ($Na_2MgAl_{10}O_{17}$) were identified in samples C2 as impurities from Al powder.

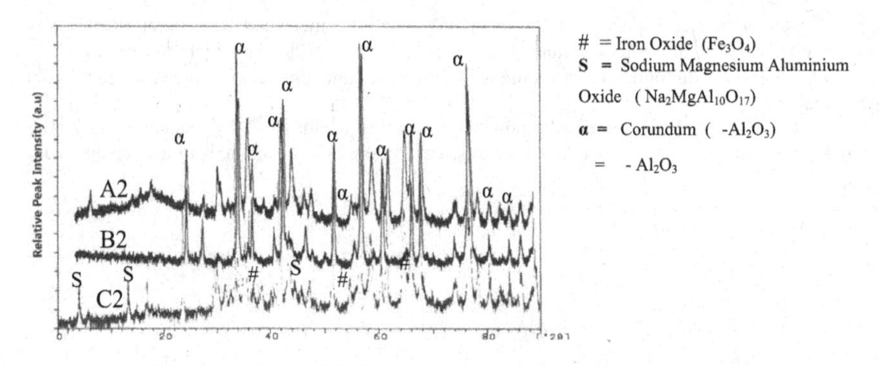

Figure 3 XRD traces taken from A2, B2, and C2 samples after sintering at 1200°C

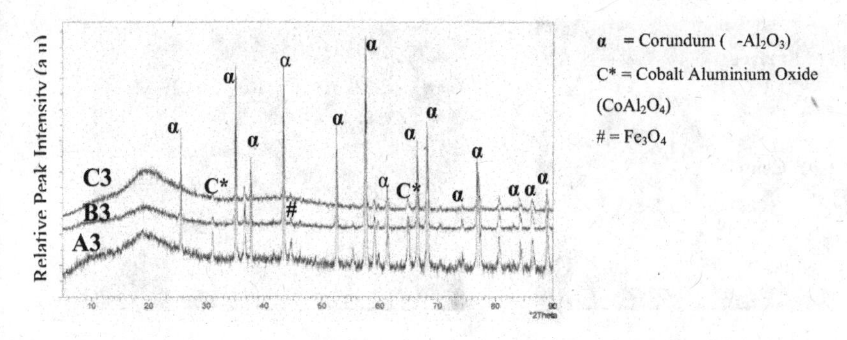

Figure 4 XRD traces taken from samples A3, B3, and C3 after sintering at 1400°C

As temperature increased to 1400°C (Figure 4), the phases present in the structure are α-Al_2O_3 (corundum) and there is no indication of transition Al_2O_3 formation. The existence of $CoAl_2O_4$ is not clear whether it was also impurities from the Al powder.

Microstructure Evolution During Heat Treatment

Figure 5 shows a part of the cross sectional microstructures from A1 samples after heating to 1000°C. Those SEM images show areas with different density across the thickness of the pellet (Figure 5a) and the laminar microstructure from the original Al flake compact (Figure 5b, d). The outer skin of the pellet appears denser (Figure 5a, c) than the inner part of the pellet (Figure 5b, d). XRD has identified the outer skin as oxide (Al_2O_3) and has been confirmed by EDAX analysis. The denser outer surface compared to the inner part results from the ease of oxidation of Al powders on the surface.The thickness of the oxide skin formed is >100-200 μm. It is interesting to find that the thickness of the dense skin formed on the vertical wall of the pellets is about twice the thickness of the top surface. Closer study on Figure 5c, some voids which are formed between layers of Al powders' stacking in the structure remained on the surface area.

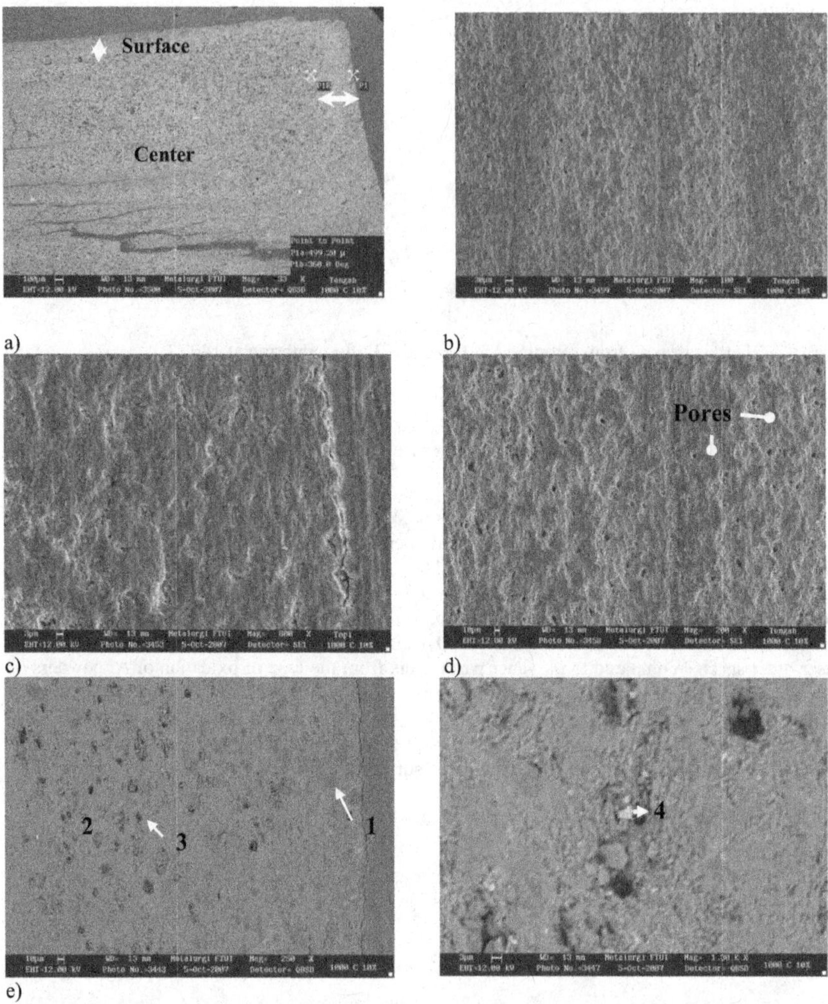

Figure 5. SEM images of A1 sample a) cross section in BSE mode, b), d) center/inner part, c) dense outer skin, e) surface and center region in BSE mode, f) white phase (number 4) identified as Fe.

Figures 5 e and f show the microstructure in BSE mode. Al_2O_3 was identified by EDAX analysis and shown as grey phase (number 1) and the pores are darker in color. Fe as impurities are also detected as bright phase (number 4). Oxidation to Al_2O_3 occurs preferentially near the sample surface or where large laminar defects (resulting from the initial compact of the Al flakes) allow easy access to the furnace atmosphere. The formation of the dense layer of α-Al_2O_3 on the surface acts as a barrier for the access of oxygen to the interior of the sample. Therefore there is a large amount of Al metal unoxidised in the sample interior as detected by EDAX (Figure 5e – number 2).

SEM cross sections of samples B1 (Figure 6a) and C1 (Figure 7a) show similar structure with sample A1 which their surface denser than the center. The thickness of the dense area of sample B1 increases with the increase of corn starch added to the Al powders. A thickness of ± 500 μm of the dense outer skin on the top surface and ± 900 μm measured on the vertical wall of the pellet. Figure 6 d-f show that the pores left by starch particles dominate, with the pores appearing well dispersed overall, both in term of distribution and separation. The pores of semispherical shape and of approximately 10 μm in size, correlating well with with the size and shape of the starch granules. Similar findings in sample B1 from EDAX analysis with sample A1 are shown in Figure 6g. Phases numbered as 1 and 3 are Al_2O_3 phase and rich-Al phase left unoxidised (number 2). Fe is also detectable.

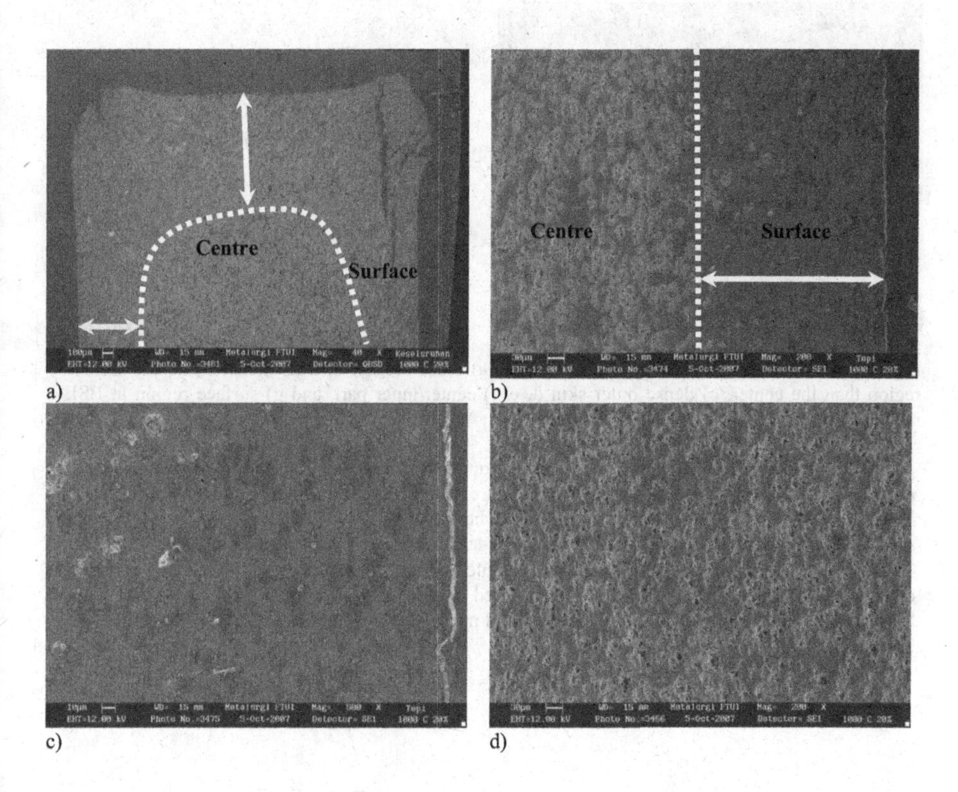

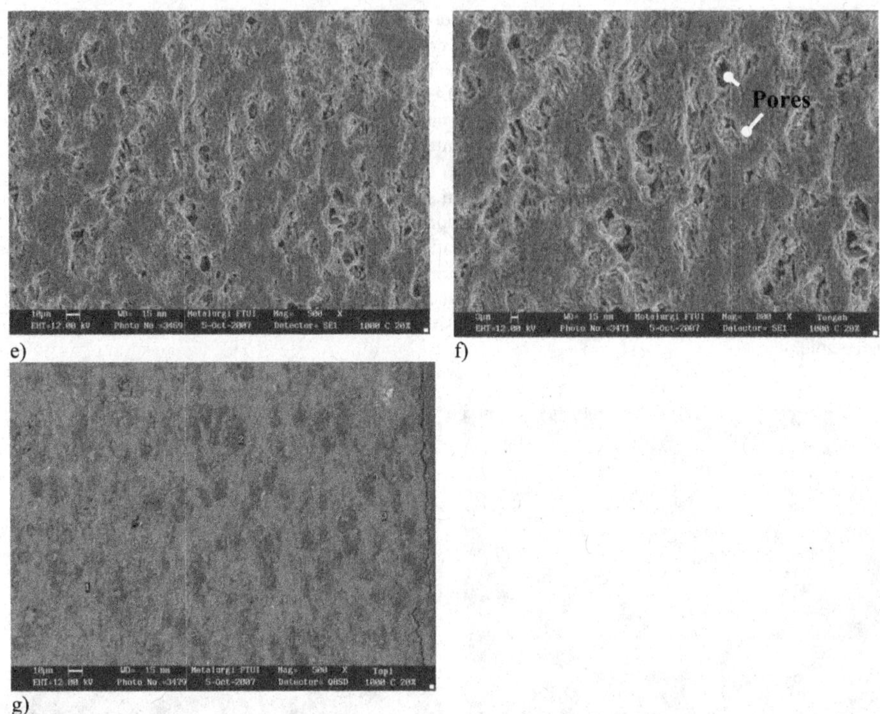

Figure 6 SEM images of *as-sintered* B1 sample a) cross section in BSE mode, b) denser surface region than the center, c) dense outer skin d, e, f) center/inner part, and g) surface region in BSE mode for EDAX analysis.

The increase of corn starch in the green compacts would increase the volume percentage of pores in the sintered body due to the completion of starch burning at temperature <400°C.[12,13] The increase number of pores in the structure eases the pathway for O_2 to access the inner part of the sample. The increase in thickness of the dense oxide skin is clearly evident of that easy access. Figure 7a shows that the dense top surface region of sample C1 has a thickness of $\pm$ 800 μm and $\pm$ 1500 μm on the vertical wall of the pellet. EDAX analysed similar phases as found in samples A1 and B1. Silicon (Si) is also found in the analysis but it did not appear as white contrast as Fe as it disolves in Al-rich phase.

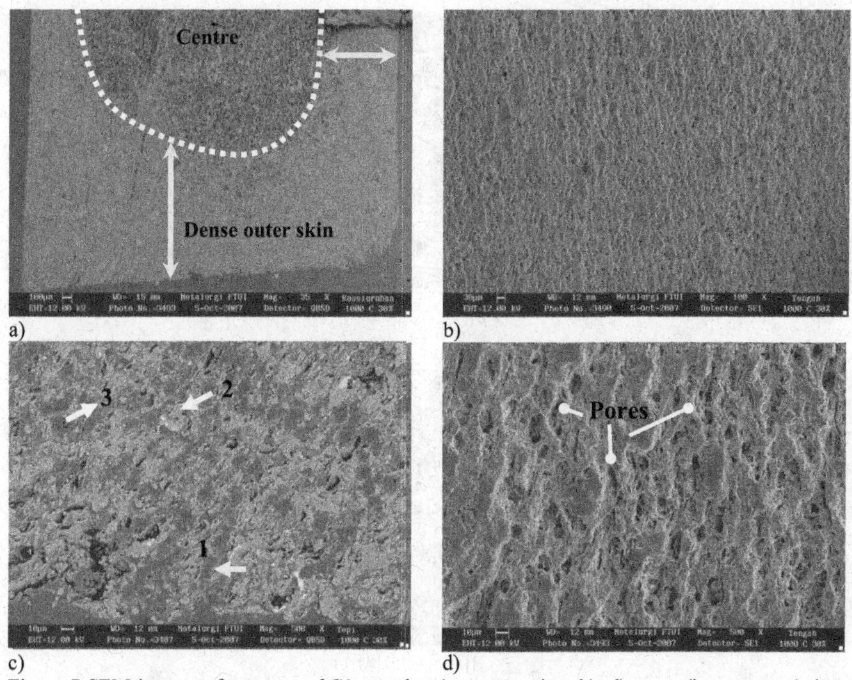

Figure 7 SEM images of *as-sintered* C1 sample a) cross section, b), d) center/inner part, c) dense outer skin in BSE mode for EDAX analysis (numbers 1 and 3 = Al_2O_3 and 2 = rich-Al phase).

After heating to 1200°C, due to barrier formed by dense oxide skin in the outer surface, it results in lowering the oxidation rate in the surface region. Figure 8c shows different density in the outer part of the samples. Figures 8b, d, f show the comparison in the number of pores and their distribution in the structure. Solid-state sintering continues to take place when samples were heated to 1400°C (Figure 9) where the size of pores remained in the structure become smaller.

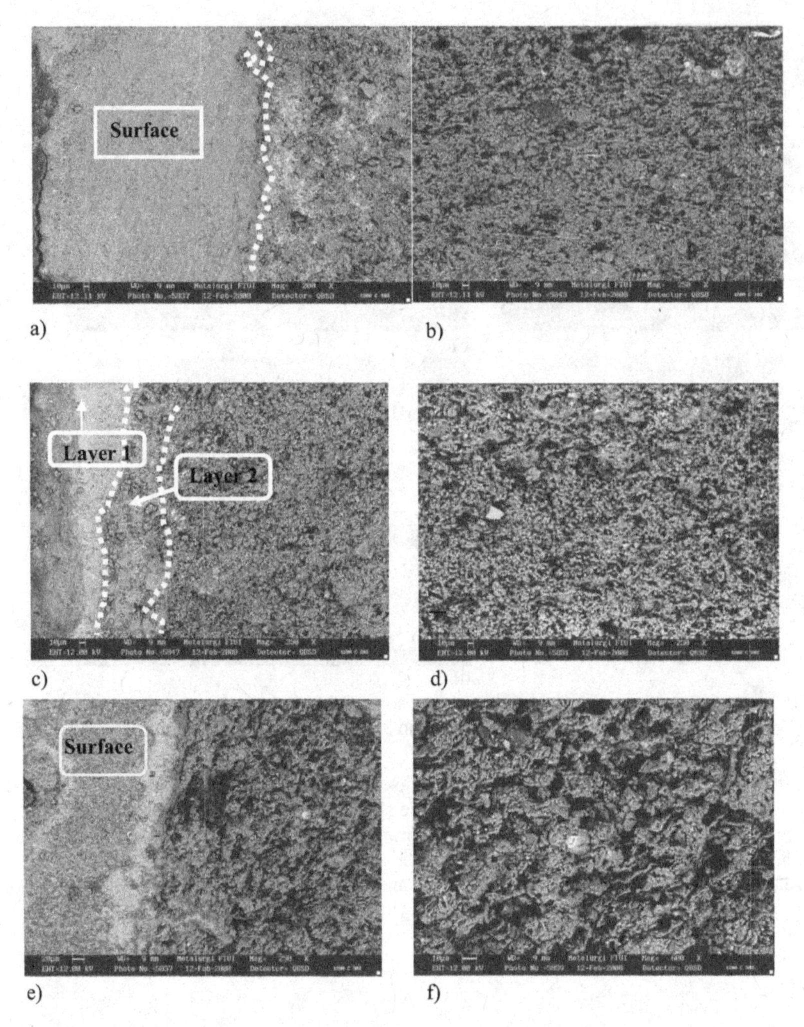

Figure 8 SEM images of *as-sintered* sample heated to 1200°C a) sample A2 – edge, b) A2 - center, c) B2-edge, d) B_2 - center, e) C_2 - edge, f) C_2 - center.

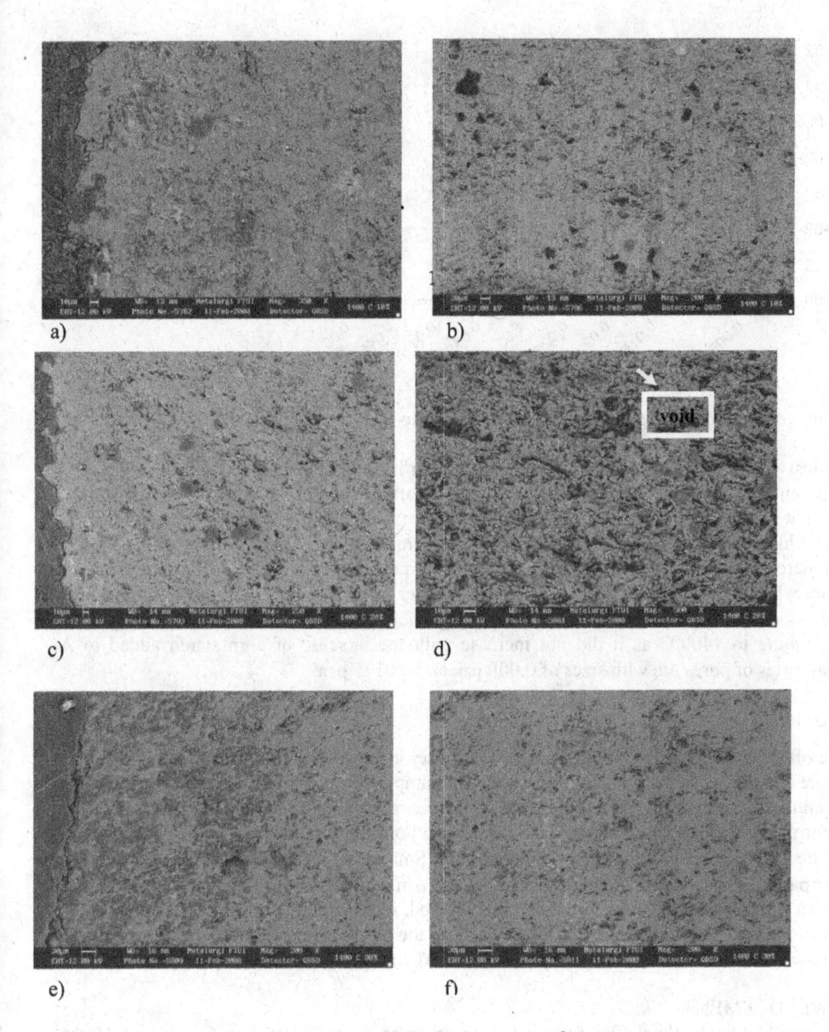

a) b) c) d) e) f)

Figure 9 SEM images of *as-sintered* sample heated to 1400°C a) sample A3 – edge, b) A3 - center, c) B3-edge, d) B3 - center, e) C3 - edge, and f) C3 - center.

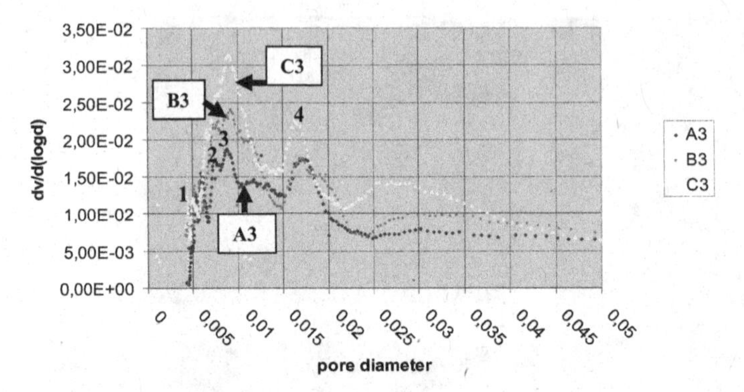

Figure 10 Pore volume frequencies of samples after heating to 1400°C

Porosimeter test was performed on samples with maximum heating temperature (1400°C). Figure 10 shows the amount and sizes of pores present in the samples after heating. The highest number of pores was identified at samples with 30 wt.% corn starch and there were identified several peaks in the plot of Figure 10 which shows 4 peaks in all samples. The increase in the amount of corn starch added results in an increase of pore volume frequencies of pores in peak 2, 3, and 4. The increase of peak 4 is not significant when the addition of corn starch to Al powder from 10 wt.% to 20 wt.%. From Figure 10, it is understood that peak 1 (0.005 μm) is the interparticle pores created from the packing of Al flakes after heating to 1400°C as it did not increase with the increase of corn starch added to Al powders. Majorities of pores are with sizes of 0.008 μm and 0.0175 μm.

4. CONCLUSIONS

The increase of wt. % of corn starch added to the Al powder increase in the number of pores remained in the structure and they ease the oxidation of Al into the samples' interior. The dense outer skin which has been identified as Al_2O_3 by XRD and EDAX increases its thickness with the increase of the amount of corn starch added and the heating temperature. Pore morphology and size distribution are detected by the shape and size of the corn starch particles. Smaller size of pores (0.005 μm) originated from the interparticles stack while the bigger pores (0.008 μm and 0.0175 μm) obtained from the sites left by the corn starch particles. Therefore using this method, microstructure of porous alumina can be tailored depending on the mechanical properties and the applications aimed. However further investigation needs to be pursue to make the oxidation of Al complete to the interior samples.

5. ACKNOWLEDGEMENT

The authors would like to thank the financial support granted by Directorate General of Higher Education of Indonesia

REFERENCES

1. Studart, A. R., *et al.*, 2006, Processing Routes to Macroporous Ceramics: A Review, *J. Am. Ceram. Soc.*, 89 (6), pp. 1771-1789.
2. Lam, D.C.C., Lange, F. F., and Evans, A.G. 1994, Mechanical Properties of Partially Dense Alumina Produced from Powder Compacts, *J. Am. Ceram. Soc.*, 77 (8), pp. 2113-17.

3. Ostrowski, T. And Rödel, J., 1999, Evolution of Mechanical Properties of Porous Alumina during Free Sintering and Hot Pressing, *J. Am. Ceram. Soc.*, 82 (11), pp. 3080-86.
4. Hirschfeld, D. A., Li, T. K., and Liu, D. M., 1996, Processing of Porous Oxide Ceramics, Key Eng. Mater., 115, pp.67-79.
5. Lyckfeldt, O. dan Ferreira, J. M. F., 1998, Proccesing of Porous Ceramics by ' Starch Consolidation', J. *Eur. Ceram. Soc.*, 18, pp. 131-40.
6. Rice, R. W., 1998, Porosity of Ceramics, pp. 539. Marcel Dekker Inc, New York.
7. Scheffler, M. and Colombo, P., Cellular Ceramics: Structure, Manufacturing, Properties and Applications. p. 645. Weinheim, Wiley-VCH, 2005.
8. Will, J. and Gauckler, L. J., 1997, Ceramic Foams as Current Collector in Solid Oxide Fuel Cells (SOFC): Electrical Conductivity and Mechanical Behaviour in Proceedings of the Fifth International Symposium on Solid Oxide Fuel Cells (SOFCV), Aachen, Germany. Edited by U. Stimming, S. C. Singhal, H. Tagawa, and W. Lehnert. The Electrochemical Society Inc, Pennington, NJ.
9. P. Colombo, 2006, Conventional and Novel Processing Methods for Cellular Ceramics, Philos. Trans. Roy. Soc. A, 364, pp. 109–24.
10. Anggono, J., 2005, The Influence of the Morphology and Size of Aluminium Powders on Their Oxidation Behaviour, Prosiding Seminar Nasional Tahunan Teknik Mesin IV 21-22 Nopember 2005, Bali.
11. Anggono, J. dan Derby, B., 2005, Intermediate Phases in Mullite Synthesis via Aluminium and Alumina Filled Polymethylsiloxane, *J. Am. Ceram. Soc.*, 88 (8), pp. 2085-2091.
12. Kumazawa *et al.*, 2004, Honeycomb Ceramics Structure Body and Method for Producing the Same, Free Patens Online Journal of Honeycomb, November 2004.
13. Anggono, J., *et al.*, 2010, Structural and Thermal Study of Al_2O_3 Produced by Oxidation of Al-Powders Mixed with Corn Starch, Ceramic Transactions volume 220: "Processing and Properties of Advanced Ceramics and Composites II," ed. Bansal, et al., pp. 299-309, John Wiley & Sons.

CHARACTERIZATION OF CERAMIC POWDERS DURING COMPACTION USING ELECTRICAL MEASUREMENTS

Timothy Pruyn and Rosario A. Gerhardt*
School of Materials Science and Engineering
Georgia Institute of Technology
Atlanta, GA, USA

ABSTRACT

In this study we evaluated the electrical response of ceramic compacts during dry pressing as a function of applied pressure. Semiconductive SiC powders were used for the experiments. In order to determine the influence of porosity in the ceramic powder compacts, a custom made die with an insulating outer sleeve was used to carry out dc and ac measurements. Measurements were performed as a function of loading and unloading compaction pressure. Dc measurements can only detect the combined response from the powders and the porosity. However, from the impedance spectroscopy data, two semicircles were observed in the complex impedance plot that allow separation of the two processes. One of these semicircles represents the bulk material property, while the other is likely due to the void space and interfaces. An estimate of the volume porosity at the different stages of compaction is provided.

INTRODUCTION

The green state of powder compacts can have a wide range of porosity values, as high as about 65%. This void space and its distribution throughout the powder compact will have a prominent effect on the overall properties of the compact.[1] In particular, it is expected that the presence of porosity will affect the dielectric constant and resistivity of the powder compacts, since porosity decreases the dielectric constant due to air having a value of 1 and increases the resistivity due to the blocking of charge carriers.[1-3] It is believed that in-situ measurements during the physical densification of powder compacts may be quite descriptive of this blocking behavior of pores in ceramic compacts, as well as be able to detect the influence of an evolving microstructure on the other dielectric properties. The measurement of the dielectric properties during physical compaction of powders by the application of mechanical pressure to ceramic powders has so far had little investigation. There is some work in the literature using diamond anvil cells at high temperatures and pressures on geological type specimens.[4-5]

For the most part; however, the study of the evolving microstructure of ceramics using impedance spectroscopy has primarily been done ex-situ after sintering. A few in-situ measurements during sintering of ceramic compacts have been carried out.[6-8] In these sintering studies, it was often observed that around the second and third stage of sintering, the grain boundary behavior is one of the primary responses observed in impedance spectroscopy and tends to dominate other blocking mechanisms such as porosity.[8]

While these studies provided a descriptive explanation of the microstructure and the electrical response after compaction, no previous studies have reported on the effect of the physical densification of powders during compaction. The present work deals with in-situ measurements of the compaction of initially loose ceramic powders at room temperature into powder compacts using a custom made die. The main objective of this work was to determine whether impedance spectroscopy was a viable method to study the compaction of powders and to determine if this could help lead to an optimization of the green state of powder compacts. In addition, this would provide another route to observing the influence of the evolving porosity and microstructure on the resultant electrical properties.

---

*Contact email: rosario.gerhardt@mse.gatech.edu

Both dc and ac measurements were utilized, but the majority of the work was conducted with impedance spectroscopy because this type of measurement can provide a great deal more information than just dc measurements.

Impedance spectroscopy is a very useful method for characterizing the electrical properties of materials since due to different relaxation times, the behavior of different regions can be separated out.[9,10] Impedance spectroscopy is also advantageous in that it is non- destructive and is one of the few options to study the porosity of materials in-situ.[2] Measurements on the effect of porosity on a wide variety of materials have been studied over the years,[1-3] but most of the work has been conducted on insulating ceramics which were dominated by the dielectric response. In this paper, we are reporting on the compaction of SiC, a semiconductive ceramic, where both the dielectric and the impedance spectra will be able to be used to monitor the influence of the porosity on the electrical properties during compaction.

EXPERIMENTAL PROCEDURE

In order to perform these experiments in-situ and under pressure, a special die with an inner diameter of 10.41mm was fabricated that involved two steel pins and an insulating teflon sleeve.[11] A schematic of the cell can be seen in Figure 1. The main concept behind this die is that the steel pins would compact the powder the same way as any other die. At the same time, leads can be attached to the top and bottom pins so that both ac and dc measurements can be made. Since the Teflon sleeve is highly insulating, a closed circuit would be made by the steel pins and the powder. In this way the electrical response of the powder could be measured as a function of the applied pressure. An insulating layer was also placed between the steel pins and the plates of the press in order to prevent any signal going through the press.

Semiconductive SiC powder, with an average particle size of about 16μm, obtained from Buehler was used for the compaction experiments. The amount of powder used for each experiment was 0.500g and was carefully placed within the die in order to ensure that all of the powder remained in between the two steel pins. The pressure was applied to the sample-cell using a Carver press and the pressure was varied from 0 up to 300 MPa. This was due to the limitations of the die, as well as these pressure values being common as compaction pressures for powders. The "loading" cycle involved applying a uniaxial pressure in segments of 10MPa and the electrical response of the pressed powder was scanned using ac and dc measurements. At the maximum pressure of 300MPa, the "unloading" cycle was started. Unloading involved releasing the pressure in segments of 10MPa and scanning the samples with both ac and dc measurements once again. This was continued until all the pressure was released. The thickness of the pellet was monitored at each pressure by measuring the initial distance between the outer edges of the steel pins and the change in length at each pressure. All of the experiments were conducted at room temperature and humidity.

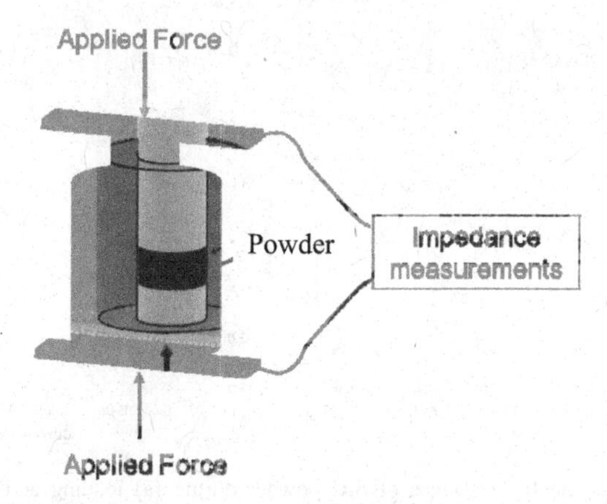

Figure 1. Schematic of the custom made die designed for in-situ impedance measurements.

The ac impedance spectroscopy scans were done using a Solartron 1296 Dielectric Interface and a 1260 Response Frequency Analyzer. The frequency was varied from 1MHz to 10mHz with the ac voltage set at a constant 500mV with no applied dc bias. Some experiments were also performed with varying ac voltage and dc bias. For the dc measurements, a Keithley 6430 sub-femtoamp remote sourcemeter was used and the voltage was measured by scanning from -100nA to 100nA. Experiments without the SiC powder showed that the impedance or resistance of the empty die was small enough, about five to six orders of magnitude smaller, that it can be neglected when compared to the ac and dc response with the SiC powder since the impedance of the powder is substantially higher than the response of the metal die alone.

RESULTS AND DISCUSSION

In order to verify that there was a proper contact between the electrode/steel pin and the powder, I-V tests were first obtained using dc measurements. It can be seen in Figure 2 that the I-V behavior was quite linear for both the loading and unloading cycles, which suggests an ohmic contact between the powder sample and the electrode/steel pins. Figure 3 shows the dc resistance at each pressure for both the loading and unloading cycles. Based on the dc results, it is clear that the compaction of the powder has a substantial effect on the overall resistance. As the pressure is increased, the size and amount of porosity is expected to decrease, which should cause the measured resistance to decrease. This is clearly observed in Figs. 2 and 3.

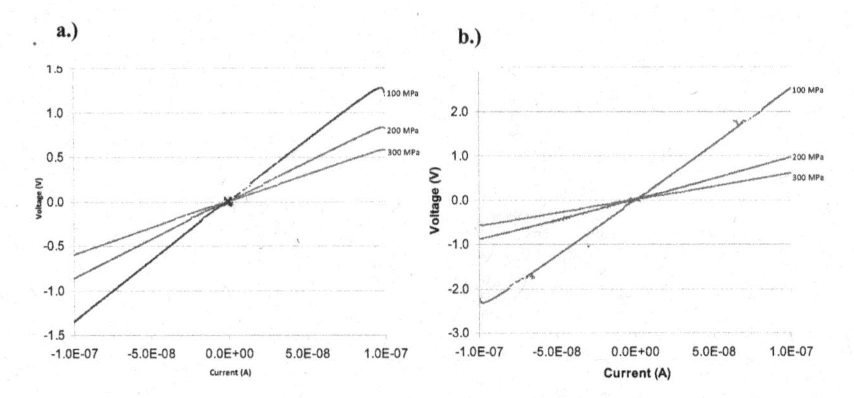

Figure 2. The I-V response of SiC powder during (a) loading and (b) unloading at three different applied pressures.

In contrast, it can be seen that the resistance during unloading behaves a little differently. At any given pressure, the unloading measurement had a smaller resistance than the loading pressure. This is believed to be a result of the powder compact not having a reason to spring back to its fully porous state during the unloading, after it has been fully compacted. Thus, the unloading dc measurement is less sensitive than the loading dc measurement.

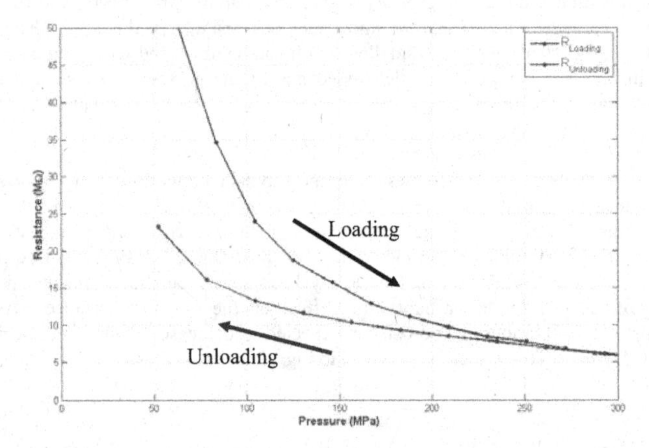

Figure 3. The measured dc resistance during compaction of SiC powders as a function of applied pressure.

A similar trend during loading and unloading can also be observed in Fig. 4, where the calculated percent porosity is shown as a function of the applied pressure during loading and unloading. The porosity is seen to decrease rapidly upon loading, whereas a much smaller change in porosity was obtained during the unloading. The porosity values shown in Fig. 4 were obtained from a rule of mixture equation shown in equation (1) below, where ρ_{comp} is the density of the compact, the theoretical density of α-SiC is ρ_{SiC}, and ρ_{air} is the density of air, V_{pore} is the volume fraction of pores, and V_{SiC} is the volume fraction of SiC. The density of the compact was found by using the mass and geometry of the pellet.

$$\rho_{Comp} = V_{pore}\rho_{air} + V_{SiC}\rho_{SiC} \qquad (1)$$

The similarities in Figs. 3 and 4 suggest that the electrical measurements provide a good qualitative indication of the porosity present in the SiC compacts. We will now move on to describe the ac measurements.

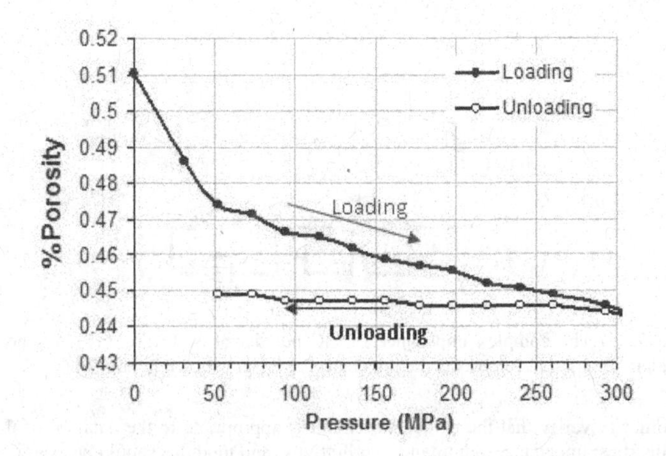

Figure 4. The percent porosity SiC compacts, as determined from the geometric bulk density, for both the loading and unloading cycles.

Figure 5 depicts a typical complex impedance plot for a SiC compact under a pressure of 100MPa. At first glance it appears that there is just one semicircle in the first quadrant. At high frequencies; however, there appears to be a partial semicircle. Examining the complex impedance plots of all the measurements taken indicate that this system can be represented by two parallel RC circuits in series. A schematic representation of this equivalent circuit is shown in Figure 5(b). Further analysis in terms of all of the dielectric functions will be presented elsewhere.

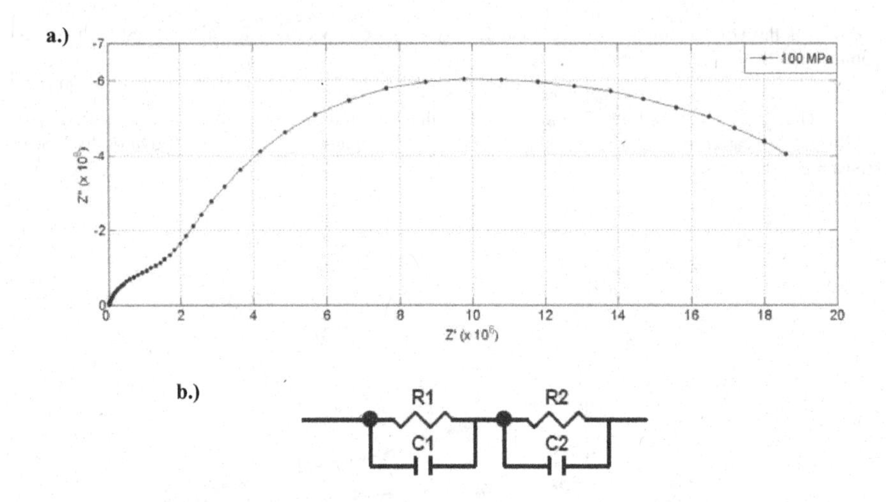

Figure 5. (a) The complex impedance of SiC powder at 100 MPa. (b) A schematic of the equivalent circuit that represents the impedance spectrum shown in part (a).

. In order to verify that the equivalent circuit is appropriate to the data, all of the frequency dependent properties: impedance, admittance, permittivity, and modulus should be examined.[12] While identifying an equivalent circuit is important, it is necessary to correlate the microstructure of the material to the equivalent circuit. This requires not only looking at the impedance, which is mostly sensitive to the conductivity but also the dielectric properties which will be more sensitive to the porosity. But first we need to understand what happens to these semicircles as a function of applied pressure during compaction. In Fig. 6(a) we can see an example of the impedance spectra for the SiC powder compacts during the loading cycle. Figure 6(b) shows the same powder during the unloading of the pressure. From these figures it is clear that the impedance semicircles quickly decrease in size as pressure is increased, as might be expected. An inset for both (a) and (b) is shown to show that the behavior of the smaller semicircle also decreases as the pressure is increased.

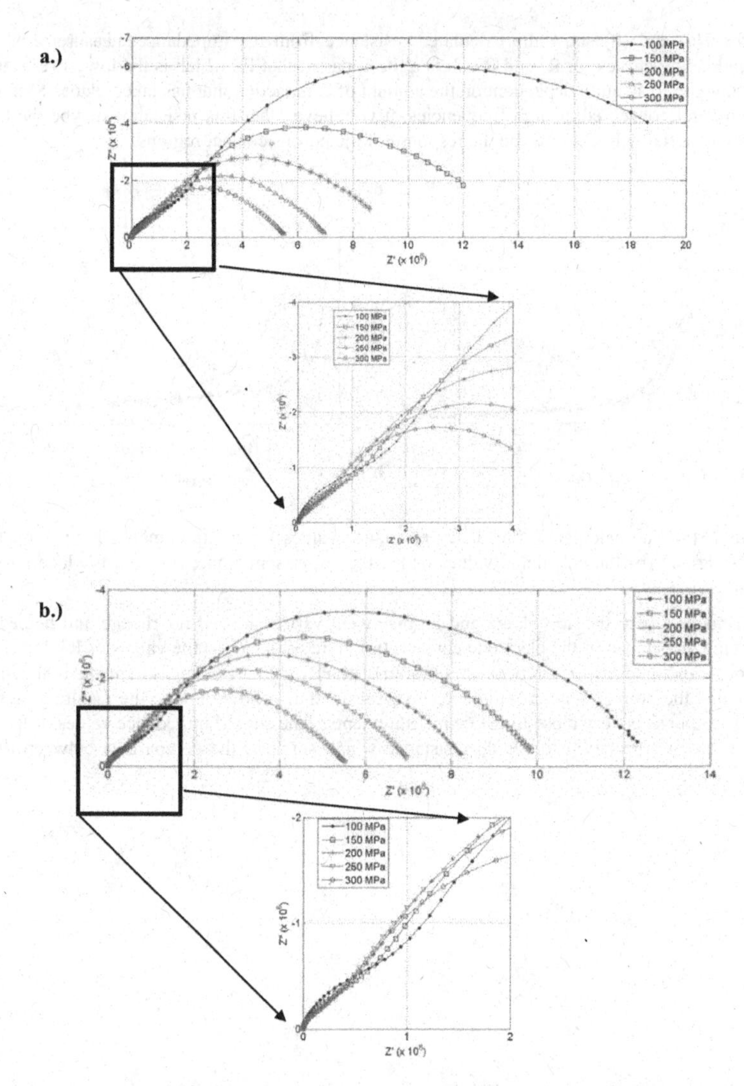

Figure 6. (a) The complex impedance of SiC powders compacted from 100MPa to 300MPa during the loading cycle. (b) The complex impedance behavior of the SiC powder form 300MPa to 100MPa during the unloading cycle. An inset was added to show the behavior of the smaller semicircle for (a) and (b).

Figures 7(a) and (b) show the calculated resistance from the impedance measurements for loading and unloading curves for R_1 and R_2. It is quite apparent that R_1, which is the lowest resistance is unchanging, while R_2 is quite dependent on the amount of compaction that has taken place. Since R_1 is also the semicircle observed at high frequencies, it is believed that this resistance maybe the bulk response of the material, while R_2 maybe the response of the porous contact regions.

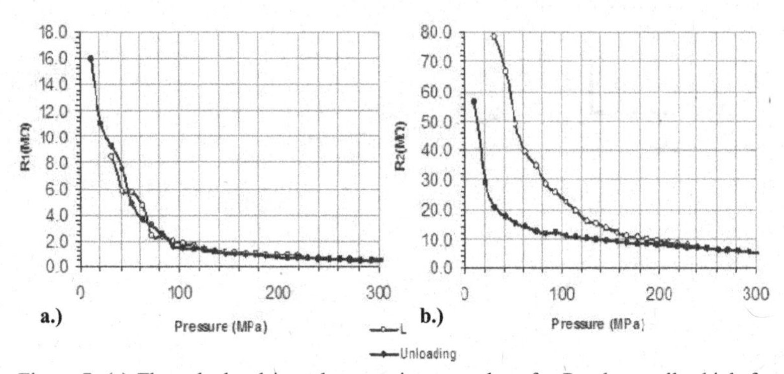

Figure 7. (a) The calculated impedance resistance values for R_1, the smaller high frequency semicircle and (b) the calculated values of R_2, the larger semicircle, for both the loading and unloading cycles.

Experiments where the ac voltage and dc bias were varied showed no change and helped us rule out R_2 being a response of the electrode contact. In Figure 8(a) and (b) the values of R_1 (bulk), R_2 (porous regions), and the comparison between the sum of R_1+R_2 with the dc measurements is shown. It is quite clear that the porous contact regions change substantially and dominate the resulting current path since this response is very close to the dc measurements. The sum of impedance values of R_1 and R_2 also has a strong correlation to the dc resistance which supports the comparison between these experiments.

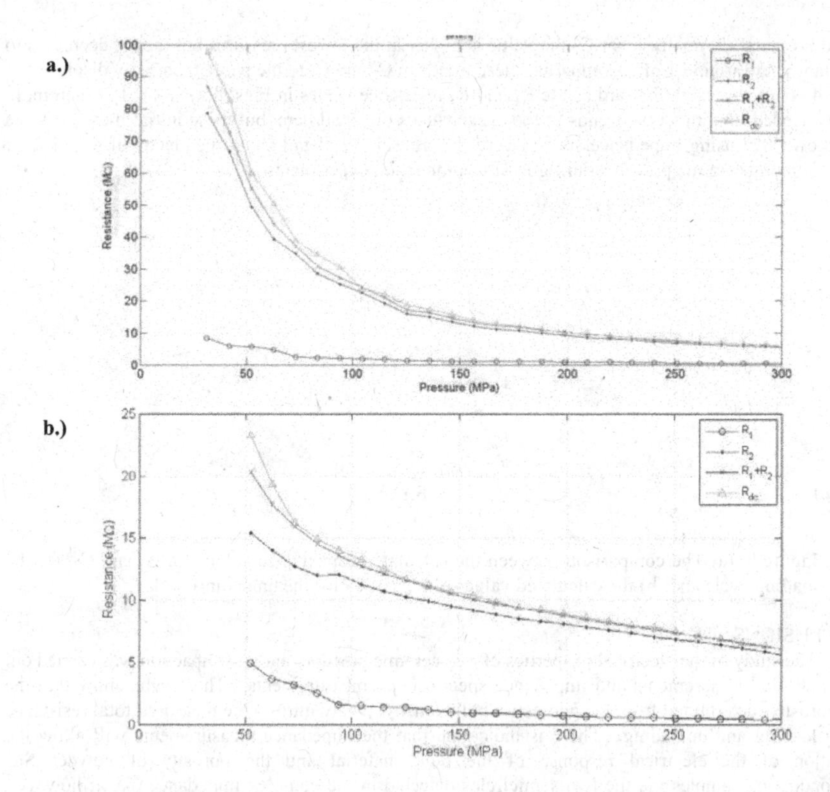

Figure 8. (a) The comparison between the calculated impedance resistance values for R_1, R_2, R_1+R_2 and the dc resistance for the loading cycle and (b) the calculated values of R_1, R_2, R_1+R_2, and the dc the dc resistance for the unloading cycle.

It is also necessary to evaluate the other circuit elements present in the equivalent circuit shown in Figure 5(b), i.e. the capacitances. The correct way to obtain the capacitance is to locate the frequency at the maximum in each complex impedance semicircle (or the frequency location in the imaginary impedance vs. frequency plot).[11] Using this value and the value of the resistance for the corresponding semicircle, the capacitance values are obtained as follows:

$$C = \frac{1}{2\pi \cdot f \cdot R} \tag{2}$$

where f is the frequency at this peak, C is the capacitance of the semicircle, and R is the resistance of the corresponding semicircle. The calculated values for C_1 and C_2 are shown in Figures 9 (a) and (b) for the loading and unloading cycles, respectively. It is clear that the capacitance C_1 is about an order of magnitude lower than the capacitance C_2 at all pressures during the loading or unloading. C_1 starts out low at first and then it increases to a relatively low constant value of ~150 pF upon increasing the applied pressure. In contrast, for C_2 the value is higher at the lowest pressures but then it decreases to a constant value around 1 nF. Comparing these trends in C_1 and C_2 to the percent porosity displayed in Figure 4 is not as straightforward as the R_1 and R_2 resistance values in Figs. 8(a) and (b). Much more work is needed to unravel the trends in the capacitances obtained here, but these initial measurements have shown that using impedance spectroscopy can reveal rich detail about the electrical response of semiconducting ceramic powders during in-situ compaction experiments.

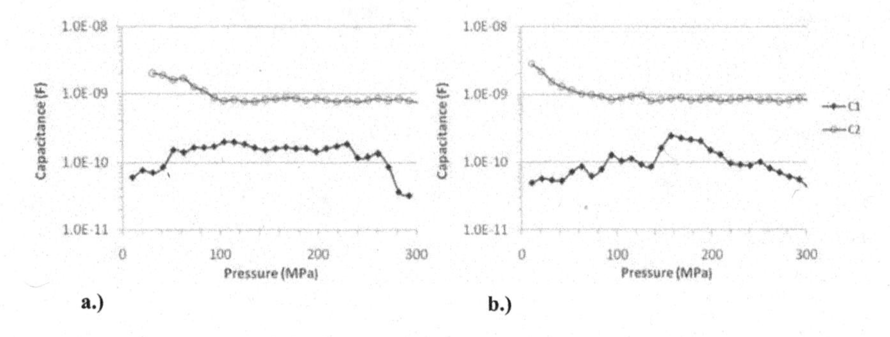

Figure 9. (a) The comparison between the calculated capacitance values for C_1 and C_2, for the loading cycle and (b) the calculated values of C_1 and C_2 for the unloading cycle.

CONCLUSIONS

The study of the electrical properties of SiC ceramic powders under compaction was carried out using both dc measurements and impedance spectroscopy measurements. The results show that the bulk porosity, determined from the geometric bulk density, nicely mirrors the measured total resistance during loading and unloading. There is indication that the impedance measurements will allow the separation of the electrical response of the bulk material and the porosity of powder SiC semiconducting samples via the two semicircles detected in the complex impedance plane; however, the capacitance values show a much more complex behavior than the resistance values that requires further study.

ACKNOWLEDGEMENTS

Funding was provided by NSF DMR-0604211.

REFERENCES

[1]W. Cao, R. Gerhardt, and J. B. Wachtman, "Low-Permittivity Porous Silica by a Colloidal Processing Method," *Advances in Ceramics,* **26** 409-18 (1989).

[2]R. Gerhardt and T. R. Grossman, "Characterization of Porosity in Thermal Barrier Coatings," *Ceramic Transactions,* **11** 189-99 (1990).

[3]K. J. Duchow and R. A. Gerhardt, "Dielectric Characterization of wood and wood infiltrated with ceramic precursors," *Materials Science and Engineering,* **C4** 125-31 (1996).

[4]C. He, C. Gao, Y. Ma, M. Li, A. Hao, X. Huang, B. Liu, D. Zhang, C. Yu, G. Zou, Y. Li, H. Li, X. Li, andJ. Liu, "In situ electrical impedance spectroscopy under high pressure on diamond anvil cell," *Applied Physics Letters,* **91** (2007).

[5]J. H. Ter Heege and J. Renner, "In situ impedance spectroscopy on pyrophyllite and CaCO3 at high pressure and temperature: Phase transformations and kinetics of atomistic transport," *Physics and Chemistry of Minerals,* **34** 445-65 (2007).
[6]D. Z. d. Florio and R. Mucillo, "Sintering of Zirconia-Yttria Ceramics Studied by Impedance-Spectroscopy," *Solid State Ionics,* **123** 301-05 (1999).
[7]F. C. Fonseca and R. Mucillo, "Impedance Spectroscopy Analysis of Percolation in (Yttria-Stabilzed Zirconiza)-Yttria Ceramic Composites," *Solid State Ionics,* **166** 157-65 (2004).
[8]M. C. Steil, F. Thevenot, and M. Kleitz, "Densification of Yttria-Stabilized Zirconia: Impedance Spectroscopy Analysis," *Journal of Electrochemical Society,* **144**[1] 390-98 (1997).
[9]I. D. Raistrick, D. R. Franceschetti, and J. R. Macdonald, "Impedance Spectroscopy: Theory, Experiment, and Applications." John Wiley & Sons, Inc: Hoboken, NJ, (2005).
[10]R.A. Gerhardt, in Encyclopedia of Condensed Matter Physics, Elsevier, pp. 350-363(2005).
[11]Charles J. Capozzi, Ph.D. Thesis, Georgia Institute of Technology, Atlanta, GA, 2009.
[12]R.A. Gerhardt, "Impedance and Dielectric Spectroscopy Revisited: Distinguishing long range conductivity from localized relaxation," *Journal of Physics and Chemistry of Solids* **55,** 1494(1994).

Author Index

An, C., 137
Anggono, J., 185

Bale, H. A., 49
Bruno, G., 137
Bucciotti, F., 95

Chen, S., 19

Dahl, R., 25
Day, D. E., 101
Duan, B., 37

Efremov, A. M., 137

Falka, M. M., 111
Fu, H., 101

Gaia, D., 153
Gerhardt, R. A., 199
Gianella, S., 153
Girija, E. K., 3
Gören, R., 177

Hanagata, N., 19
Hanan, J. C., 49
Hayakawa, S., 13
Huang, T. S., 65
Hsu, H.-P., 25

Ikoma, T., 19

Jaina, R. H., 111
Jainb, H., 111
Johnson, B. Y., 123

Kalkura, S. K., 3
Kearney, C., 25
Kowala, T. J., 111
Krevolin, J., 25
Kumar, G. S., 3

Li, J., 13, 19
Liu, J. J., 25
Liu, X., 65, 123
Lu, W. W., 37

Madinelli, A., 95
Mandel, S., 79
Manocha, L. M., 163
Manocha, S., 163
Marzilliera, J. Y., 111
Mert, I., 79
Morita, H., 19

Nickerson, S., 137

Ortona, A., 153
Osaka, A., 13, 19

Owusu, M. O., 123
Özgür, C., 177

Patel, H., 163
Patel, K., 25
Piccinini, M., 95
Pruyn, T., 199

Rahaman, M. N., 65, 101
Ruffin, M. K., 123

Şan, O., 177
Sglavo, V. M., 95
Shavitri, I. A. O. R. S., 185
Shirosaki, Y., 13
Spector, M., 25

Takeguchi, M., 19
Tamura, N., 49
Tas, A. C., 79
Thamizhavel, A., 3
Tjitro, S., 185

Wallen, A., 25
Wang, M., 37
Wangb, S., 111

Yokogawa, Y., 3